$55.50

Paper Contracting

The How-To of Construction Management Contracting

by
William D. Mitchell & Gary W. Moselle

- Turn your estimate into a bid.
- Turn your bid into a contract.
- ConstructionContractWriter.com

Craftsman Book Company
6058 Corte del Cedro / P.O. Box 6500 / Carlsbad, CA 92018

The cartoonists featured in this book are available for freelance work and may be contacted directly:

Theresa McCracken: www.mchumor.com
Jim Whiting: www.jimtoons.com

Looking for other construction reference manuals?
Craftsman has the books to fill your needs. **Call toll-free 1-800-829-8123** or write to Craftsman Book Company, P.O. Box 6500, Carlsbad, CA 92018 for a **FREE CATALOG** of over 100 books, including how-to manuals, annual cost books, and estimating software.
Visit our Website: www.craftsman-book.com

Library of Congress Cataloging-in-Publication Data

Moselle, Gary.
 Paper contracting : how to make a living as a construction management contractor / by Gary W. Moselle and William D. Mitchell.
   p. cm.
 Includes index.
 ISBN 978-1-57218-270-7
 1. Construction contracts--United States. 2. Construction industry--Management--United States. 3. Contracting out--United States. 4. Contractors--Legal status, laws, etc.--United States. I. Mitchell, William D. II. Title.
KF902.M67 2012
690.068'4--dc23
                    2011052110

©2012 Craftsman Book Company
Layout: Joan Hamilton/Christine Bruneau, Cover design by Tradebookandmedia.com

Contents

1 Paper Contracting — 5
- Why Owners Choose Construction Managers 8
- Construction Management Summarized 9
- Opportunities for CM Contractors 11
- What's Different about CM Contracting? 12
- Overhead & Profit 16
- The CM Contractor's Fee 17

2 A Case Study — 23
- The Facts 24
- Lessons for CM Contractors 31

3 Managing Pre-Construction Services — 37
- A Contractor's Perspective 37
- Adding Value During the Design Phase 38
- Finding Prospects 40
- The Initial Contact: What's Required? .. 41
- The Second Meeting: Make an Offer 47
- The Third Meeting: Get a Commitment 48
- A Formula for Success 50
- Q & A on CM Contracts 55
- Your Design & Engineering Team 57
- The Design Process 61
- How Architects Work 62
- Your Contract for Pre-Construction Services 64

4 Review the Plans and Specs — 67
- Plan Checking 68
- The Six Cs of Plan Review 70
- Cost Estimates 77
- Pulling the Permit 79
- The First Schedule 82

5 Preparing Bid Packages & Evaluating Bids — 87
- Public vs. Private Contracts 88
- The Bid Package 89
- Getting the Word Out 90
- The Invitation to Bid 95
- The Pre-Bid Conference 97
- Awarding the Contract 100
- Invitation to Bid Form 103

6 Reviewing the Trade Contracts — 111
- Essentials in Every Construction Contract 112
- Optional Contract Terms 114
- Contract Bias 116
- Comply with State Law 118
- Your Own Contract 120
- Ordering Materials and Services 121

7 Observing Day-to-Day Construction — 125

- Direct, Oversee, Monitor and Advise 126
- Bonds and Insurance 127
- Staying on Schedule 131
- Contractor Claims for Delay 140
- LEED Green Projects 144
- Jobsite Safety 145
- Defective Work 147
- Quality Assurance 148
- Hazardous Materials 149
- Cleanup 149

8 Keeping the Owner Informed — 153

- Heading Off Problems 154
- Information Channels 156
- Submittals, Samples, Shop Drawings 159
- Significance of Approval 162
- Retaining Construction Records 163
- Your Job Log 165
- Contacts for Everyone 167
- Record Documents 168

9 Evaluating Payment Requests — 171

- The Request for Payment 172
- The Schedule of Values 175
- Final Payment 177
- Release of Retainage 179
- Interest on Past Due Balances 181
- Pay or Reject? 182
- Rejections 183
- Liens & Waivers 187
- Keeping Track of Payments 189
- Face Time with Owner 192

10 Communicating with Contractors and Suppliers — 195

- Limits of Authority 197
- Manufacturer's Instructions 198
- Interpreting the Contract 199
- Interpreting the Plans & Specs 201
- Correcting Design Defects 205
- Job Conferences 207
- Suspension of the Job 209
- Termination for Cause 211

11 Assisting with Change Orders — 213

- Mutual Agreement vs. Force Account 215
- Negotiating Price Changes 216
- Processing Change Orders 221
- Value Engineering 224

12 Protecting Against Construction Claims — 227

- Allocating Risk of Loss 228
- Primary Source of Claims: Surprises 232
- Site Walk 233
- Errors in the Plans 235
- Notice of Discrepancy 236
- Differing Site Conditions 236
- Resolving Contractor Claims 239
- Mediation, Arbitration & Litigation 241

13 Directing Project Closeout — 247

- Beneficial Occupancy 247
- Inspection for Substantial Completion 249
- Substantial Completion Punch List 254
- Final Completion 256
- Callbacks 257
- Express and Implied Warranties 258
- Warranty Exclusions 261
- Responsibility for Warranty Claims 262

Index 265

Chapter 1

Paper Contracting

YOU'VE HEARD THE TERM *paper contractor*. Such as, "That guy is only a paper contractor." But what does that term really *mean* in the construction community?

The term, as most of us in construction understand it, is used to distinguish between the *typical* general contractors (those with payroll and specialized equipment) and those who only manage the construction process and don't actually do the work (those *without* payroll and specialized equipment).

The implication on the street is that anyone with a business card could become a paper contractor overnight. Even worse, the term implies no acceptance of responsibility — that a paper contractor has nothing to lose. If the job goes bad, he (or she) has no skin in the game. They could simply walk away, leaving others to deal with the mess.

If that's what paper contracting means to you, keep reading. There may be reasons to take another look.

Here's our definition of *paper contracting*:

> ➢ A paper contractor has a contractual agreement with a property owner to manage the design, budgeting, scheduling, bidding and construction of a specific project.

- The paper contractor agrees to manage that project in exchange for a specific contractual fee.
- This fee may include a financial incentive for controlling costs.
- The contract may or may not include a guaranteed maximum price (GMP).
- The trade contracts are prepared by the paper contractor; however, it's the owner, not the paper contractor, who actually signs contracts with trade contractors.
- The paper contractor may advise on pre-construction issues such as design and cost control.
- The paper contractor will make recommendations on all the issues normally handled by a general contractor: selection of trade contractors, scheduling, approval of payments, resolving claims, and project closeout.

So, as far as being responsible for the complete construction process is concerned, a paper contractor operates under a legal contract just as any typical contractor would. The difference is that a paper contractor escapes the obligations associated with having employees on staff and all of the headaches of owning, maintaining and insuring expensive equipment.

Is anyone really doing this?

The answer is yes. In fact, it's the most common form of construction contract for state and local government. Many of the largest projects undertaken by the Corps of Engineers are now done on this basis. Many small commercial and residential jobs are also done by paper contractors. Some of the most experienced, most successful, most respected construction professionals we know fit the definition of paper contractor.

For example, we know of a paper contractor who makes a very good living doing residential work almost exclusively. His clientele is very comfortable dealing with consultants — lawyers, accountants, financial advisors, etc. In their eyes, he's just another consultant, a *construction* consultant. This contractor gets involved early in the project, before design has started. He recommends an architect and an engineer, gets owner approval of the plans, guarantees a maximum price, selects trade contractors, ensures project completion to the owner's satisfaction and usually collects a nice premium for completing work on time and under budget. It's a good business. But it's not the typical general contracting business as most people understand it across the nation. It's a paper contracting business that embodies far less risk than is normally taken on by typical general contractors.

Is this legal?

Of course it is. No state requires that general contractors have employees and payroll. In fact, the potential profit deck is stacked against any general contractor that supports large crews: worker's compensation insurance, high overhead, high payroll taxes and liability insurance premiums.

But please be aware of this: We're not going to recommend anything that's even marginally illegal. Be assured, everything you read in this manual is strictly legitimate. If you're looking for a scheme to avoid taxes or bend the rules or for a way to flim-flam clients, please look elsewhere. We stand foursquare behind high professional standards and ethical conduct for all types of construction contractors. We assume you won't tolerate anything else.

Is this just another construction management book?

No, it's a paper contractor how-to book, and all a paper contractor does is manage. This is an important distinction that needs to be clear right up front. There are in this business, *general contractors, construction managers, program managers* and *paper contractors*.

General contractors have employees and specialized equipment.

Construction managers generally don't have many field employees and very little, if any, specialized equipment. A construction manager is usually doing the construction *at risk*, meaning they're on the hook financially for the entire construction process, including the labor and materials.

A *program manager* quite simply manages the program, that is, the designers and contractors. But they don't manage the day-to-day construction.

A *paper contractor,* by way of distinction, manages the day-to-day construction, but never has any field employees, nor does he have any equipment beyond a phone, a laptop computer, and a calculator.

Now, we assume you're already experienced at getting work done on a construction site, so this isn't a book about how to build a project; it's a book about how to manage a construction project with the least amount of risk, the least amount of staff, and no equipment. What we're going to show you here is a different way of doing business contractually. We're going to provide the tools and the background you need to adapt to an ever-changing environment for construction services.

The Way It Was

Fifty years ago it was unusual to find any construction project that wasn't headed by a conventional general contractor. The lowest responsible bidder got the job. Much of the work (such as excavating, concrete, framing and drywall) was done by crews employed by the general contractor. The remainder of the construction work was done by contractor specialists (subcontractors) working under agreements with the prime contractor. Invoices flowed up the chain of contractors to the owner and money flowed down the same chain from the owner or lender to subcontractors and suppliers. The prime contractor ran the show, handling invoices, claims, disputes, payments, change orders and other requests.

General contracting as it was practiced 50 years ago isn't going to vanish. But today, more projects, especially major projects, are handled by construction companies that do little or none of the actual construction. Paper construction management has become a specialty with its own set of rules, standards and qualifications.

Why Owners Choose Construction Managers

Most property owners know very little about design and construction: finding qualified design personnel, preparing bid documents, researching potential contractors, evaluating bids, responding to change orders and requests for information. Very few property owners want to manage the day-to-day operation of a construction project.

A construction manager is the owner's representative — answers only to the owner. The primary role of that manager is to protect the owner against excessive cost, delay, poor craftsmanship and unnecessary risk. Everyone else on a construction site has some other agenda.

Becoming a Paper Contractor

You probably picked up this book because you know plenty about construction contracting already. You may have made a living as a general contractor. We hope you've completed at least a few projects successfully and, in the process, started building a reputation for quality work and professionalism. We hope property owners come calling when they need a bid on a proposed project. We hope you enjoy what you're doing — the smell of fresh-cut lumber and the sight of construction materials coming together to better serve the people who will live and work in what you're creating.

"When I grow up I want to be a CM contractor like my daddy so I can blow my top at subcontractors like he does."

If you're the general contractor we've just described, you almost certainly recognize problems associated with traditional general contracting: risk of loss, disputes, excessive regulation, employee problems, warranty claims, change orders, callbacks and, perhaps most difficult, the constant need for additional working capital. Traditional general contracting is fraught with enormous risk-taking, and it doesn't take many things going wrong to put you under. But if you're a paper contractor, your risks are minimal.

It would be a shame if headaches like those drove you out of construction. What you do best and what you enjoy most is bringing the parts together to create something of value. What you don't like are the headaches that nag general contractors — and eventually drive many into another line of work. If you're looking for a better way to make a living in the construction industry — a way that makes good use of what you know about construction — please keep reading. That's precisely what we're going to explain.

Paper Contracting, aka Construction Management Contracting (CM for short)

We've elected to title this book *Paper Contracting*. We did that for several reasons. First, the term is both descriptive and widely understood. Second, the title emphasizes the intellectual part of construction, not the physical process. What you're selling is your skill and experience, the same as any consultant (attorney, accountant, financial advisor, etc.). Third, the term *paper contracting* is more descriptive and better understood than any of the alternatives, such as *construction consultant*, or *construction management contractor* or *owner's representative*.

But we recognize that some, including some prospective clients, will be offended by the term. If you put "Paper Contractor" on your business card, it's likely that you'll make a poor first impression among prospective clients. So we need a better description when referring to the way you do business. After some soul searching, we settled on *CM contracting* or *CM* for short. That's the term we're going to use for the remainder of this book.

CM Summarized

To restate: a construction management contract is an agreement to pay for consulting services. The construction manager oversees project development in exchange for a fee. The project owner, not the construction manager, will sign agreements with trade contractors. The construction manager may advise on pre-construction issues such as design and cost control and will make recommendations on contract awards, coordination of trades, purchase of materials, payment of trade contractors, scheduling, and project closeout. The sidebar on the next page has a list of tasks usually handled by CM contractors.

As you read down this list, you'll probably say to yourself, "That's what I'm doing now" and that's precisely the point! That's what every general contractor does. But it's what a CM contractor *doesn't do* that better defines the process.

> **The job description for a CM contractor will usually include the following tasks:**
>
> - review of the plans and specifications
> - preparation of bid packages and evaluation of bids
> - reviewing owner and trade insurance coverage
> - approval of the proposed trade contracts
> - communication with designers, trade contractors and suppliers
> - overseeing the building permit process
> - monitoring the day-to-day construction process
> - keeping the owner informed of the design and construction progress
> - review and approval of trade and suppliers' invoices
> - assistance with budgets, schedules and change orders
> - protection of the owner from construction claims
> - directing project equipment testing, warrantees and closeout

What's *not* included in a construction manager's portfolio? That's easy. A CM contractor earns a fee for consulting services — period. The CM doesn't buy or install materials. A CM has no contracts with the trades and pays no bills. The owner signs agreements with trade contractors, pays all the bills and ultimately holds the installing contractors responsible for their work.

A CM contractor earns a fee as a consultant, the same as any architect or engineer. That fee could be a set amount for the job or a set amount per month, week or hour. Or it could be a percentage of the job cost. Or the fee could be based on performance — sometimes, depending on the contractual agreement, in which case it would be called a *CM at risk* contract. We'll explore each of these options later.

CM Contracting by Type of Job

The type of job will usually influence where you fit in the jobsite organizational chart. The charts following illustrate a CM's role on three of the most common types of construction projects. The arrows show the flow of information, requests, suggestions, submittals and approvals. These arrows are *not* lines of authority (to give instructions) or liability (for charges), as we'll explain later.

- *Residential project* — The CM assumes many responsibilities normally handled by a general contractor.
- *Commercial project* — The CM acts as a buffer between the owner and the general contractor.
- *Tenant improvement project* — One CM may represent the owner. Another CM may represent the tenant.

OPPORTUNITIES FOR CM CONTRACTORS

Residential

Home improvement and new home construction projects are good candidates for CM. These jobs are small enough so that no prime contractor is required. Instead, what often happens is that building departments allow homeowners to act as their own general contractor, which is known in the trade as an Owner Builder project. However, few owners are experienced in construction management. Most owners aren't prepared to participate in the design process either. Under these conditions a CM contractor can help deliver a superior product that meets the owner's expectations.

Residential Project
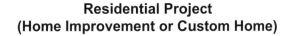
(Home Improvement or Custom Home)

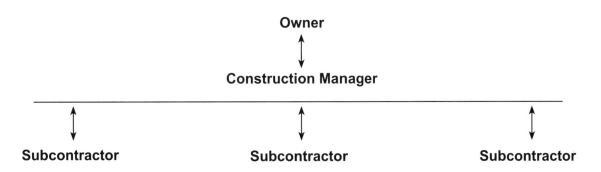

Commercial

These projects tend to be larger in scope. There may be several prime contractors on a large commercial job. One prime trade/contractor may finish before others begin work. The CM contractor provides continuity from beginning to end. The work of a CM begins with design and site requirements (port-o-pots, temporary fence, job trailer, etc.) and ends at final cleanup and project closeout. The CM has to work closely with each prime trade/contractor to be sure the project remains on schedule and under budget.

Commercial Project
(Shopping Center or Office)

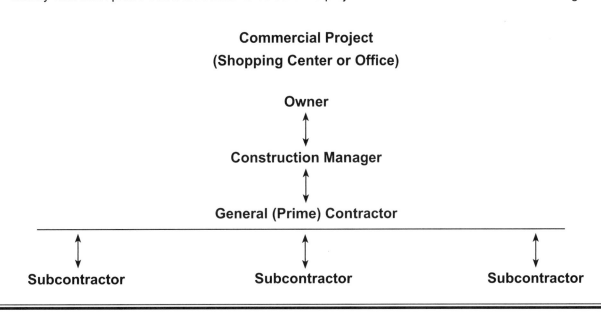

Tenant Improvements

A larger commercial tenant improvement project may have two CM contractors. The building owner who is funding the project will be represented by one CM contractor. The building tenant who will occupy the completed space may be represented by another CM contractor. Each CM contractor will have distinct interests and goals. The role of both CM contractors will be to meet the specific project expectations of each party.

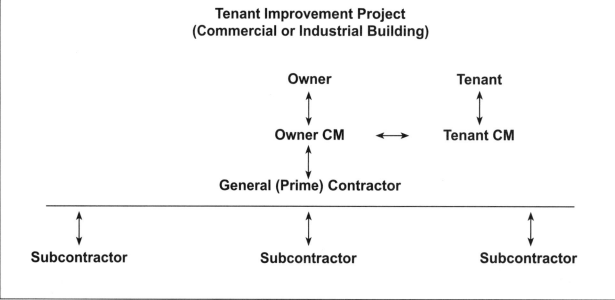

Regardless of the type of job, all CM contracts have two elements in common.

1. The CM contract binds only the owner and the CM contractor, not the actual construction.
2. The primary responsibility of the CM is always to protect the interest of the owner.

What's Different About CM Contracting?

Most of a CM contractor's task will come as second nature to an experienced general contractor. After all, CM contracting is still construction, whether public works or private construction. The project could be a new commercial or residential building, non-building construction, tenant improvements or home improvements. But that's where the similarities end. There are many differences between CM contracting and traditional general contracting.

The Contract

CM contracts are very different from traditional prime contracts, and have to be drawn carefully to avoid problems with property owners, trade contractors, suppliers and state regulators. Because CM contracting is so different from traditional general contracting, it's easy for owners and others to be confused. That's why your CM contract is so important. In the next chapter, we'll offer a case study in what can go wrong when an attempt at CM contracting falls off track. For now, just understand that a good CM contract draws a line in the sand, placing limits on both the authority and the responsibility of a CM contractor.

> *"Because CM contracting is so different from traditional general contracting, it's easy for owners and others to be confused."*

When you do CM work for a public agency, that agency will almost certainly prepare the contract. Your only responsibility may be to sign the agreement. If you build or remodel homes, offices or commercial buildings or handle industrial projects, you'll usually have the opportunity to offer a CM contract as part of the negotiating process.

Later we'll offer a link on the Web to a selection of blank CM contracts and trade contracts. We'll also suggest a way to get free CM contracts for any of the 50 states or the District of Columbia.

Recommending vs. Giving Orders

If you've been working as a traditional general contractor, you may not feel comfortable on your first CM job. A general contractor gives instructions to subs and suppliers and expects those instructions to be followed. A general contractor usually has written agreements with subs and can enforce penalties against subs who don't comply. That's not true with a CM (consulting) contract. Only the owner has written agreements with trade contractors. The owner may have many agreements with trade contractors on a single project. Those agreements are usually referred to as *multi-prime* contracts. Typically, the trade contractors are recruited by the CM contractor, are almost certainly approved by the CM contractor, and will work closely with the CM during construction. But these trade contractors are legally responsible only to the owner. The CM can recommend and encourage, but the owner makes the decisions. You, as the owner's CM, only have to see that those decisions are carried out.

Obviously, this difference in CM contracting cuts both ways. True, the CM's authority is limited, and that's typically bad, at least from the CM's perspective. But the upside to that limited authority is that it also brings with it a certain amount of immunity, in that the CM isn't legally responsible for the failure of a trade contractor to perform. Of course, a CM can get a black eye when there's trouble with a trade contractor, which is something you'll have to guard against. But there's no legal liability. The CM doesn't have to make good on a trade contractor's mistakes. That's between the owner and trade contractor, and is up to them to settle. You're just the owner's advisor.

Paying Bills

A CM contractor earns a single fee as a consultant, and that's the only real cash that passes through his hands. All other money that constitutes payments for work completed on a project, after the CM's review, goes straight from the owner to trade contractors and suppliers. Again, this cuts two ways. A CM contractor doesn't have the benefit of float that results from holding (for a few days) payments made for work by subcontractors. But neither does a CM have the temptation to divert payments to some purpose other than paying subs. Some states impose a trust on payments for subcontract work held by a general contractor. Diverting those payments to any purpose other than what's intended would be a violation of that trust. In some states, that carries criminal penalties.

Licensing

If your state requires that contractors be licensed, do you need a license to work as a CM contractor? That's a very good question. Before we answer, remember that CM contractors don't install materials and don't have subcontractors. So, do CMs have to be licensed?

Unfortunately, this is a question that remains unsettled in many states. Generally, states that require contractors to be licensed define very precisely the term *construction contractor*. Sometimes the definition includes construction management and sometimes it doesn't. For example, Washington D.C. Code of Municipal Regulations § 17-3905.1 to § 17-3905.14 requires that construction managers comply with all standards that apply to general contractors.

In Tennessee, CM contractors need a license (Lowrey v. Tritan Group Ltd., 2009 U.S. Dist. LEXIS 60312). The same is true in New York (Liberty Management & Construction v. Wasserman, 1996 U.S. Dist. LEXIS 4408).

California courts have decided (at least for now) that CM contractors don't need a contracting license. If you're interested, the case is The Fifth Day, LLC v. Bolotin, 172 Cal. App. 4th 939. The case was decided in March of 2009 and includes a well-reasoned dissent by Justice Mosk. Expect the Bolotin decision to be reversed the first time a homeowner brings suit against an unlicensed CM home improvement contractor.

In Nevada, homeowners can act as their own contractor. However, if you're a general contractor acting as the owner's representative, the state licensing board will tend to view you as the responsible party if a dispute arises on the project.

In most other states where contractors have to be licensed, the licensing of CM contractors remains an open question. You should check with your local contractor licensing board before beginning any actual work.

For the purposes of this book, we're going to assume that CM contractors need the same license that a general contractor would need if working on the same job. We do this for three reasons:

1. Having the appropriate license is reassuring to everyone concerned — the owner, trade contractors, suppliers, lenders and design professionals.
2. We believe courts and legislatures are going to recognize the growing popularity of CM contracting and close any loophole that permits CM contractors to work without a license. You don't want to be caught in that loophole just when the loophole gets plugged.
3. Finally, the penalty for operating without a license is severe. In many jurisdictions, an unlicensed contractor has no right to collect. Don't take that risk — nobody likes to work for nothing.

Warranty vs. No Warranty

A hundred years ago, the rule in construction was *caveat emptor*, buyer beware. If the basement leaked or the roof blew off or if the foundation settled, the contractor wasn't responsible — unless the contract included an express written warranty. Gradually that's changed — in either or both of two ways. First, courts started assuming an *implied* (unwritten) warranty of workmanlike performance in every construction contract. Anything that didn't meet that standard was a breach of the court's implied warranty. The contractor had to fix the problem or pay for repairs. How long that warranty of workmanlike construction lasted and what it covered had to be settled on a case-by-case basis. Worse, warranty law was different in every state because construction is governed by state rather than federal law, and each state has its own rules.

Later, state legislatures jumped in to help protect consumers. Some states passed laws that made contractors liable for specific warranty repairs for a specific period. Other states protected consumers with their license law, threatening to revoke the license of any contractor who refused to repair shoddy work. Other states required coverage with warranty insurance. Still others set up mediation procedures intended to resolve warranty claims out of court.

Now, where does a CM contractor stand in this warranty muddle? There's no way to be sure. CM contracting hasn't been around long enough. We don't know of any CM contractor who has had to answer a warranty claim in court. It simply hasn't happened yet. Until it does, we can only guess. And here's our best guess.

Court-made warranty law (*workmanlike construction*) probably doesn't expose CM contractors to any risk. A CM contractor doesn't install construction materials, and so, is not a workman. That's the easy part.

But what about warranties imposed by statute? Can a custom home builder avoid all risk of warranty claim by working as a CM contractor? Stated that way, we feel the answer has to be "No." If the house slides off its foundation, everyone on the job is going to be named as a defendant — the engineer, the excavator, the cement mason, the concrete supplier, the framer and the CM contractor. Working as a CM doesn't provide total immunity from warranty claims. But CM contractor status clearly tilts the playing field in your favor. Remember, the owner of that sinking home approved and paid the engineer, the excavator, the cement mason, the concrete supplier and the framer as well as the CM contractor. Among those, the least likely to be found liable is the CM contractor. Of course, a job well done is your best protection against future claims, but it's not a guarantee.

"Were the building contractors of the Leaning Tower of Pisa sued?"

O&P

Every general contractor's bid price includes markup — overhead and profit. Most bids also include a small contingency allowance. So isn't a contractor who signs a CM agreement giving away the potential for profit? The answer has to be a qualified "maybe."

Let's suppose you've received an offer to bid on a particular job that's right up your alley. Trained crews are on hand and available. There are very few unknowns. You've got plenty of working capital and bonding capacity available. You've done work like this many times, maybe for the same owner. And you've made good money every time. Is that job a candidate for a CM contract? Probably not. If you know exactly what you're doing and have no doubt about how the job will work out, go ahead and bid the work on a fixed fee basis.

Now let's reverse the assumptions. The type of work is a little different from what you've done before. There are multiple unknowns. Finding trained crew members could be a problem. Working capital and bonding capacity may fall short of what's needed. The owner could be hard to work with and may be undercapitalized. The plans seem a little primitive. There are sure to be changes once work starts. Is that job a candidate for your CM contract? *You bet it is,* because it's filled with risk, and doing the job on a CM basis goes a long way to vaccinate you against those risks.

But aren't you giving away the profit on that job? Not at all. There's more risk than profit in a job like that. Taking the job on a CM contract eliminates most of that risk. As a CM, you're paid for your time. If your contract contains project time limits that, if exceeded, would allow you to rework your fee, the inevitable delays on a job like this only mean you have a paid job for longer. Another benefit: if the owner runs out of cash and the job is shut down with liens, your contracting license isn't in jeopardy. As a consultant, you're not exposed to claims by trade contractors. We don't recommend taking on any job that you expect will fail. But where the risk exceeds the potential rewards, whip out your CM contract.

In short, CM contacting isn't for every job. Instead, choose wisely. Opt for a CM contract when it suits both your needs and the needs of an owner. Make the pitch, "I can save you some money on this job. I'll act as your consultant at a price my competition can't touch. There's zero profit for me in this job. Just pay me for my time. I'll give your job the same attention I give to every project I take on, but with no markup. You can't beat that deal. Say the word and I'll write up a contract we both can live with."

A CM Contractor's Fee

Traditional construction contracting came in two flavors — either (1) fixed price or (2) time and materials (cost-plus). CM contractors have many more choices, from simple to complex. Choices are described below in the order of most simple to more complex.

Hourly rate

Everyone can understand this. Lawyers usually bill by the hour. CM contractors can too. *The advantage:* Clients should receive regular statements showing the time worked and a short note about what was completed in each time period. If there's a dispute about any item billed, the amount in question is likely to be small. Presumably, each statement will be paid before the next statement is sent. So there shouldn't be any surprises. *The disadvantage:* Some clients may wonder if you're really putting in all that time. There won't be a time clock available where you're working. And, most of all, there's no limit to an hourly fee. The longer the job drags on, the more fees accumulate. There's no financial incentive to finish the job quickly. The solution is to put a cap on the total fee payable. The fee for the entire job can't exceed a certain amount. That can reassure an owner who's sensitive on the issue of cost. But a fee cap can also reduce an owner's incentive to make good use of your time. So, you'll need to be diligent when it comes to keeping the job on schedule or your hourly rate will drop.

Note that a cap on the CM fee is not the same as a guaranteed maximum price (GMP). We'll cover that later.

"The CM's fee is a set amount no matter how much time is required and no matter how long the job lasts."

Lump Sum

The CM's fee is a set amount no matter how much time is required and no matter how long the job lasts. This fee plan will work best on small jobs that are likely to be completed in a month or two. If the job will last many months, there will be an issue of how much is due on each statement date. A CM contractor has a financial incentive to front-load charges (larger payments at first). An owner will prefer back-loaded charges (higher payments due on completion). Another disadvantage is that if a lump sum project takes longer than expected, you can end up working for free for the period beyond what you anticipated and based your fee on.

Weekly or Monthly Charge

For example, if the job lasts 40 weeks and the rate is $2,500 per week, the total charge would be 40 times $2,500, or $100,000. Whether paid by the week or month, have a clear understanding about pro-rating any month or week before work starts, after work ends and during periods when work stops. It's common to have jobs delayed due to bad weather, material or labor shortages, code problems, or funding issues. Your fee agreement should identify what happens during an extended delay. *The advantages:* Compensation on a weekly or monthly basis will be a good choice when a CM contractor works substantially fulltime on a single job from start to completion. This is very similar to working on salary as an employee. *The disadvantages:* Not every job needs a fulltime CM. On a smaller job, the CM may be required only a few hours a week. But again, the customer will see there's no incentive for you to complete work on time or within budget.

Percentage of Construction Cost

This is where fee agreements begin to get more complex. Selecting the percentage is the easy part. Five to 10 percent is common on a smaller job such

"Or we could do it on a time-and-materials basis."

as a residence. On a large job, such as an airport or power plant, the CM contractor's fee could be as little as one or two percent. When a CM contract is offered for public bid, CM contractors may compete on the basis of their percentage. The contract will usually be awarded to the CM contractor with the lowest percentage bid.

It's important to identify very precisely what's included in the definition of "construction cost." The list of what's included and what's excluded should be very complete — resolving every possible issue, including all change order costs. Your CM contracts should include very precise definitions of construction costs for each of four possible cost categories:

➢ labor

➢ materials

➢ equipment

➢ subcontracted services

Usually the CM contractor's fee will be a percentage of all four cost categories. But in a rare case, the agreement may exclude one or more of these categories from the definition of construction costs.

CM contracts routinely exclude from the definition of construction cost:

➢ compensation of design and engineering professionals

➢ the cost of buying or owning land or rights-of-way

➢ financing costs

➢ legal fees

➢ court costs or the cost of arbitration

➢ charges imposed by government authority on the owner of the construction site

➢ the CM contractor's fee

The CM contractor usually sends an invoice at the end of each pay period based on payments made to contractors and suppliers during that pay period. The most common pay period is monthly, although it can be done by phases such as: grading, concrete, or framing.

One word of caution. Any time the CM contractor's fee is based on the construction cost, the CM contractor needs the right to audit the owner's cost records. The CM contractor won't necessarily see all invoices for work done and materials supplied. To make sure *all* costs are considered when calculating the fee, the CM contractor needs the right to inspect and copy the record of construction costs paid by the owner.

Guaranteed Maximum Price (GMP)

To this point in our discussion of CM fees, we've considered only the consulting fee. We haven't considered the construction cost. But any of the fee agreements considered so far (hourly, lump sum, weekly, monthly or percent of job cost) could include a GMP. This is usually called an *at risk CM contract*, or CM@R. The CM shares in any savings if costs are less than the GMP. But the CM may have to pay some portion of the difference if there's a cost overrun. For example, if the division is 80-20, the CM contractor would receive 20 percent of any savings below the GMP. The same split could be used to share in any cost overrun. The owner might pay 80 percent and the CM contractor might have to contribute 20 percent. Obviously, the CM has a heavy financial incentive to deliver the completed project for less than the GMP.

CM@R contracting is a potential win-win-win for contractors. First, because it's a CM contract, most of a general contractor's worst headaches are shifted to the owner: disputes about payment, liens, warranty, contractor claims, change orders, code compliance, etc. Second, because it's a CM contract, you're going to get paid for your time. Your fee is guaranteed. Third, even though it's a consulting contract, you can earn more than the consulting fee. A CM@R contract sets the maximum cost to the owner (excluding change orders) and provides a formula for division of any savings below that amount.

Another advantage of a GMP contract: Six states currently require that contracts for home improvement work show the actual cost of work to the owner.

- *California* — Business and Professions Code § 7159(d)(5)
- *Illinois* — Compiled Statutes Title 815, § 513/15
- *Massachusetts* — General Laws 142A, § 2(a)(5)
- *Nevada* (residential pools only) — Administrative Code § 624.6958-2(f)
- *Pennsylvania* — Statutes Title 73, § 517.7(a)(8)
- *Tennessee* — Code Annotated § 62-6-508(a)(5)

A CM contract based on an hourly rate or even a lump sum doesn't qualify under these state laws because there's no set amount for the work. A CM contract with a GMP qualifies under the law in all six states because there's a set maximum price for the work. In these six states, the CM contractor can share in any savings on a home improvement job if the CM also agrees to absorb any cost overrun (exclusive of change orders).

The website: http://PaperContracting.com includes sample CM@R contracts for both residential and commercial construction. The download is free.

Where We're Headed

Obviously, this chapter has been a sales pitch. If you haven't offered your services as a CM before, we hope this discussion has convinced you to give it a try. We've also used this chapter to define essential terms. What follows is your CM toolbox — the practical tools you need to make a good living as a CM contractor. We'll start with a horror story, a CM job done exactly wrong.

Managing construction is taming chaos. A successful CM consultant needs good ways to monitor and control project development. The same control system should do "what if" analysis to anticipate problems on their way to becoming expensive mistakes. We're going to call this system the CM contractor's *control loop*. The objective is to provide sound recommendations to the owner on critical construction issues: time, costs, resources, cash flow and scope of the work. If a project is on track to bankrupt an owner, someone should warn the owner as soon as possible. That's a CM's job, as you'll see in the twelve chapters that follow.

Chapter 2

A Case Study

WE SUGGESTED IN THE last chapter that CM contractors have an obligation to prevent confusion among subs, suppliers and owners. CM contracting is consulting, not general contracting. It's easy for subs, suppliers and even the owner to be confused about a CM contractor's authority and responsibility. Leave room for confusion and you risk an expensive mistake.

This chapter illustrates many of the most common mistakes a CM contractor can make. This is a true story. It happened in Suffolk County, New York. The court's decision in October of 2010 found the contractor, Joseph P. Marcario, Sr., jointly liable for $59,391.51 for building materials used to construct a 6,000 square foot home for Don Nenninger.

As you read through this case, make mental notes about what doesn't seem quite right to you. See how close your list of mistakes comes to matching our list, which appears near the end of the chapter.

We're going to quote from the court's opinion, paraphrase some parts of the opinion and add observations relevant to CM contracting. If you want to read the entire case, search on the Web for Thurber Lumber Co. Inc. v. Marcario.

The contractor, Joe Marcario, had good instincts, at least most of the time. He was trying to run the Nenninger job as a CM contractor — a consultant. He was also working on the same job as the excavation contractor under the name Mastic Beach Excavation. It seems obvious to us that Joe Marcario didn't intend to serve as general contractor on the Nenninger job. But, as we'll see, circumstances backed Joe Marcario into that role. To walk the fine line between excavation contractor and CM contractor, Marcario needed a good contract. Unfortunately, he didn't have any written agreement with Nenninger. That's always a mistake. But it wasn't Marcario's only mistake, as the court's opinion makes clear.

> **Thurber Lumber Co. Inc. v. Marcario**
> (2010 NY Slip Op 51909U)
>
> **Principal characters:**
> Joe Marcario, contractor.
> Thurber Lumber Company, supplier.
> Don Nenninger, homeowner
>
> **The place:**
> Suffolk County, New York
>
> **The time:**
> October 2003 to August 2004
> Case decided on October 26, 2010

The Facts

The court's opinion is in italics exactly as it appears in the official reports and begins with the testimony of Pagan, a salesman for Thurber Lumber Company.

Marcario went to Thurber before October 2003, with blueprints for Nenninger's new home. Thurber agreed to provide the lumber and materials which would be delivered and paid for on Marcario's account. For each delivery of materials, Thurber immediately invoiced the shipment and faxed the invoice to Marcario. Marcario had a C.O.D. account. That is, he would pay for each shipment made; he did not have a line of credit other than what would accrue as Thurber delivered the additional materials.

Observation: Notice Pagan's statement that materials "would be delivered and paid for on Marcario's account." That's not what Marcario intended on the Nenninger job, as we'll soon see. But it's what Pagan quite reasonably assumed. Marcario had been buying materials for his own account from Thurber Lumber for over 20 years. Marcario will testify that he planned to have Nenninger pay Thurber directly for each delivery, exactly as you would expect under a CM contract.

Marcario's mistake was that he didn't explain his plan for the job. Marcario intended that his involvement in the Nenninger job be as a consultant. Marcario wanted Nenninger to be liable to Thurber Lumber. Invoices could be faxed to Marcario. There's no problem with that. But those invoices should have been charges to Nenninger. Marcario should have made that point clear to Pagan from the first meeting. After that first discussion, Marcario should have sent a follow-up letter to Thurber confirming the conversation.

Would Thurber have objected to billing Nenninger for deliveries? Probably not. The law in every state gives building material dealers a lien for the value of materials delivered to a construction site. As a secured creditor, Thurber didn't need to rely on the credit of Marcario.

The first "drop" of materials to the site was made in October 2003. It was immediately invoiced and faxed to Marcario. Apparently, at Marcario's request, Nenninger, paid $16,282.00 directly to Thurber for that first "drop." Pagan testified that, subsequently, Nenninger made payments directly to Thurber for the materials invoiced to Marcario. The total of the checks paid by Nenninger was about $37,180.00. Pagan stated that when the accruing balance increased, Nenninger informed Thurber that he was applying for additional financing so that he could make further payments for the materials.

Observation: This is the first hint of trouble. Nenninger is running short of cash. He's applied for additional financing. Written contracts wouldn't be necessary if every construction project went as planned and if every owner and contractor had a perfect memory. Until that happens, you need a good contract.

Plaintiff's rebuttal witness was called upon to verify Nenninger's representation. Pagan described that he repeatedly reached out to Marcario for payment, even speaking with Marcario's secretary. The secretary confirmed receipt of the invoices. However, Marcario never returned the calls to Pagan. Marcario did, however, make payments in October 2003 to Thurber for about $15,064.00. That payment was for a special order of steel to be used by Marcario in building the new home.

Observation: Pagan still doesn't understand that Marcario is working in dual roles on the Nenninger job: First as a trade contractor (for the building foundation) and second as CM contractor (for the balance of the job). The $15,064 payment was for work on the foundation. Marcario recognized that he was liable to Thurber for steel ordered as a trade contractor. Marcario didn't plan to be liable for other materials Thurber delivered to the Nenninger job. Apparently Marcario never made that clear to Pagan.

When Marcario provided Thurber with the blueprints, Marcario informed Thurber that he authorized Phil or Matt to place all framing orders connected with the construction. Pagan stated that Nenninger never placed framing orders himself; all the orders were placed by Phil or Matt just as Marcario had authorized.

Observation: This is an issue for every CM contractor. Suppliers are accustomed to taking orders from general contractors — and holding those general contractors liable for payment. Subcontractors (trade contractors) aren't likely to be confused when dealing with a CM contractor. That's because most trade contractors will have a written contract with the property owner, not the CM contractor. But most suppliers won't have a written contract for delivery of materials. Every supplier needs to understand that the property owner, not the CM contractor, is responsible for payment on delivered materials.

Marcario testified next. He initially described himself as only an excavator. That is, his original agreement with Nenninger was to clear the property, excavate, provide the foundation, install the septic system, and set the steel framing. This work was accomplished through his company Mastic Beach Excavation. He acknowledged ordering steel from Thurber, for which he paid in advance. He also acknowledged that his Thurber account was on a C.O.D. basis and that he always paid cash for his materials. Such was his 20+ years business relationship with Thurber.

Observation: Marcario has described accurately his work as a trade contractor on the Nenninger job. Next, he's going to describe working as a CM contractor on the same job.

Although Marcario delivered the blueprints to Thurber, he denied negotiating any deals on Nenninger's behalf. He only put Nenninger in touch with Pagan, and all negotiating was between Pagan and Nenninger.

Observation: Marcario has described accurately his role as a CM contractor on the Nenninger job. If the court believes this testimony, Marcario won't be held liable for the $59,391.51 Thurber is claiming.

Significantly, Marcario initially denied having any knowledge that the construction materials were being delivered and billed under his name. Then, as his testimony progressed he admitted receiving the faxes from Thurber, indicating with his fingers almost an inch-thick of faxes he received. He never returned the invoices to Thurber, nor did he dispute the quality or quantity of the materials shipped under the invoices.

Observation: This is key testimony. If invoices had been addressed to Nenninger, Thurber would have a hard time proving that Marcario was responsible for payment. Marcario had an obligation to advise Thurber that Nenninger would be responsible

> **Statutory Employer Defined**
>
> This is an old concept, dating back to early in the 20th century, when state legislatures wrote the first worker's compensation laws. Before these laws existed, employees injured on the job had to sue their employer for compensation. Recovery could take years and often left employees with no medical treatment and without the care required for rehab. Worker's comp laws took away an employee's right to sue an employer and gave back compensation for all industrial injuries regardless of fault. Nearly all employers, including construction contractors and subcontractors, were required to buy worker's comp insurance covering employees. If a sub didn't have worker's comp coverage, the law made the next contractor up the chain, usually the prime contractor, the employer for purposes of the worker's comp statute. So a prime contractor who fails to verify the worker's comp coverage of subcontractors can become a statutory employer for worker's comp purposes.

for payment. Why didn't Marcario correct Thurber's misunderstanding? We can only guess. One possible reason is that Marcario didn't anticipate any problem. If Nenninger had enough cash to complete the job and remained on good terms with Marcario, Thurber would be paid in full. But that's a risk that no CM contractor wants to accept.

Marcario described how the subcontractors were hired by him. He explained that the subcontractors didn't have the required worker's compensation and liability coverages. By working through Mastic Beach Excavation, the subcontractors were covered under that company's worker's compensation and liability policies.

Observation: Marcario is right. Employees of a subcontractor who doesn't carry worker's compensation insurance will be covered under the general contractor's policy. Mastic Beach Excavation became what is known as the *statutory employer* of the subcontractors' workers. Mastic Beach would be responsible for worker's comp premiums due on wages paid to employees of those subcontractors. The court's opinion doesn't reveal whether Mastic Beach actually paid those premiums.

But notice the inconsistency here. A CM contractor doesn't have employees working on the job and thus doesn't have to carry worker's comp insurance on employees. Marcario's testimony about subcontract employees was an admission that he was working as a general contractor on the Nenninger job.

As construction progressed, Marcario accepted money from Nenninger to pay the subcontractors.

Observation: Marcario started working on the Nenninger job as an excavation contractor. When pitching the job to Thurber, Marcario intended to work as a CM contractor. Now Marcario has admitted operating on the Nenninger job as a general contractor. We hope you see a problem inherent in claiming all three roles. A CM contractor is protected by his or her status as a consultant.

But that protection is abandoned when a CM contractor accepts the responsibilities of a traditional general contractor.

At one point, he said, Nenninger stopped paying him, so Marcario told the subcontractors to work directly with Nenninger. The first time that Marcario indicated to Thurber that he was not liable for the invoices was when he received the final August 2004 invoice for $59, 391.51. He wrote upon it that he was not responsible, then mailed it to Thurber and Nenninger.

Observation: By August of 2004, it was obvious that the Nenninger job had gone bad. There was going to be trouble, probably legal trouble. Marcario was eager to climb back behind the protection he would have as a CM contractor. That's like trying to put Humpty Dumpty back together again.

Marcario's cross-examination provided scant additional testimony. The Mastic Beach Excavation company no longer exists; it was owned by him, his wife, and son. He acknowledged that he solicited the bids from and hired the subcontractors. Although no contract existed between him and Nenninger, he said the plan was for the two to sit down after construction and discuss what Nenninger would owe Marcario for his services.

Observation: We hope it's obvious to every experienced contractor reading these words that there's a wrong time and a right time to agree on payment for services. The wrong time is when the job is done. The right time is before work begins. That's the purpose of a signed contract. Toward the end of this chapter we'll have more to say about payment for services under a CM contract.

Nenninger testified last. He explained going to Home Depot and 84-Lumber for lumber estimates before meeting Marcario. Marcario told Nenninger that he had an account at Thurber, and if Nenninger ordered from Thurber, Marcario would pass the contractor discounts on to Nenninger.

Observation: This could be the reason why Marcario allowed Thurber to bill Marcario for materials used on the Nenninger job. A better strategy for Marcario: Explain that he, Marcario, was about to make Nenninger a very good customer at Thurber and ask that Thurber extend Marcario's discount to Nenninger.

Nenninger understood that Marcario's personal work was to excavate, arrange the foundation, arrange for plumbing and electrical, and erect the steel. Once the functioning shell was completed, Marcario would then hire subcontractors for the work inside. Nenninger stated that he initially gave $10,500 cash to Marcario to pay the subcontractors, but the subcontractors never appeared. He had to hire his own workers.

Observation: Both Nenninger and Thurber were confused about Marcario's role on this job. Anyone working as a CM contractor has an obligation to eliminate confusion about authority, liability and responsibility. The best and quickest way to do that is with a written contract.

Through the months, Nenninger denied that he ever received Thurber invoices from Marcario for payment. He did admit receiving the invoices from Thurber when Thurber could not reach Marcario.

Observation: Is this an inconsistency in Nenninger's testimony? How could Nenninger have written checks for thousands of dollars payable to Thurber and deny that he ever received an invoice from Thurber?

Nenninger acknowledged that there were no written contracts between him and Marcario. A businessman, engaged as a mortgage broker at the time, he acknowledged that he never received a receipt from Marcario for almost $50,000.00 in cash payments he made to Marcario. That $50,000.00 was in addition to any payments made by check. He intimated he may have received one or two receipts for the cash, but he was unable to produce them for trial.

Observation: Is it good practice to accept cash in payment for construction services? Banks have an obligation to accept cash. Contractors don't. Can you rely on an owner who insists on paying in cash for construction services? In this case, Judge Tarantino found Nenninger's testimony not worthy of credit.

The only testimony found credible by the Court is that provided by Thurber. The court finds that Marcario's and Nenninger's testimony was incredible. It's unbelievable that a mortgage broker would engage a contractor without a written contract to build a new home. Similarly, it's unbelievable that a contractor would construct a new home without a contract. The Court is compelled, in light of Marcario's testimony, to invoke the maxim "falsus in uno, falsus in omnibus" [false in one thing, false in everything] *regarding the totality of Marcario's testimony.*

Thurber is entitled to judgment under "account stated" as against both Marcario and Nenninger, and for goods sold and delivered, but yet unpaid. Marcario testified that he did receive and retain, for several months, the more than 80 invoices as each delivery was made. He admittedly did not refute any of the invoices as to the quantity or quality of materials delivered, nor that the account was in his personal name, nor that the materials shouldn't have been billed to his account. It was only the final demand for payment that Marcario contested, but only after months of retaining the individual invoices and accepting deliveries. Nenninger, similarly, received and retained the invoices. He also made partial payments towards the balances due Thurber for the materials used to construct the home in which he resides.

The Court also finds that Nenninger's testimony was incredible. As a real estate broker, Nenninger should have been familiar with new home construction, the protections offered by written contracts, and the implications of paying large sums of cash to a contractor without obtaining a receipt. Nenninger claimed to have paid Marcario more than $50,000.00 in cash bills, in addition to checks he paid to Marcario. Yet, not one receipt for the cash paid, nor copies of checks made, were introduced into evidence.

Observation: Next, Judge Tarantino rules on Marcario's claim that he was working as a CM contractor on the Nenninger job.

Marcario's testimony was equally incredible. With more than 20 years in the industry he proceeded with Nenninger on this project without a written agreement between him and Nenninger. Additionally, although he testified that he was only the excavator, the evidence revealed more. By his own testimony, Marcario admitted to negotiating the contracts with the subcontractors. The Nu Frame Contracting proposal revealed that, through Marcario, Nu Frame was to "build a six thousand square foot house as pertaining to plans." Moreover, on November 3, 2003, Marcario provided Nenninger with a "work completed to date" bill reflecting $16,362.13 for lumber, and $25,000.00 "for labor on framing and sheathing the house." Marcario's denial of liability for the Thurber materials which were to be paid by Nenninger is belied by the statement seeking payment from Nenninger for those materials. The Court is not impressed with what appears to be a scheme where the contractor tells the homeowner to pay the lumberyard directly to create the appearance that the contractor is not liable for its account to the lumberyard.

Observation: Re-read that last sentence of Judge Tarantino's opinion. As explained in Chapter 1, nothing in the law prevents a contractor from working as a consultant. The judge isn't saying that it can't be done. In fact, it happens every day. Judge Tarantino is saying that it can't be done by an agreement between the contractor and the property owner alone. We agree. And that's the point of this chapter. Every CM contractor has to make it clear. He or she is working as a consultant. Other contractors, subcontractors, suppliers and service providers on the job have to understand that. The CM contractor's agreement is for consulting services alone.

The Order of the Court

Judge Tarantino found both Marcario and Nenninger liable for the full amount Thurber claimed, $59,391.51, plus interest from August 31, 2004. Marcario was prohibited from trying to enforce any contract he may have had with Nenninger. The court's order gave Thurber the right to collect the full amount from either Marcario or Nenninger.

Lessons for CM Contractors

Some of Joe Marcario's mistakes are obvious. Others may not be so apparent. Here's our list. We hope you were able to identify most of these mistakes.

1. *Have a written CM contract with the property owner.* The contract should make it clear that you're working as a consultant, not as a general contractor. The contract should state specifically that you're not going to order or install materials. The website: http://PaperContracting.com includes CM contracts that meet these standards.

    Everyone wants a deal, especially an owner spending a boatload of money on a new house. But asking a subcontractor to manage a large construction project without a contract is like skydiving without a parachute. Expect disaster. Subcontractors should pick a role and stick to it. As a CM contractor, you have an opportunity to lay off nearly all construction risk. Don't muff that chance by ordering materials or employing anybody ... period!

2. *Be sure trade contractors know they're working for the property owner.* The easiest way to do this is with a written contract. Send trade contractors a memo defining your position. You're a consultant. Trade contractors work 100 percent for the owner. You're not in any way responsible.

3. *Be sure suppliers understand that the property owner will be liable for material deliveries.* Reject promptly any invoice addressed to anyone other than the property owner. If necessary, get a credit application from key supply houses. Have the owner apply for accounts where needed. An owner with enough credit to build a big house will qualify for credit at any supply yard I know. If the owner doesn't qualify for a line of credit, maybe you've got an even bigger problem than you realize.

4. *It's OK to talk directly with trade contractors.* That's an important part of every CM contractor's job, as we'll point out in the chapters that follow. But resist the temptation to give instructions. Instructions have to come from the owner.

"Don't worry. The first 20 years in this business are the hardest."

This may sound a bit confusing if you've been making a living as a general contractor. How can you build anything without giving instruction? Easy. The same way a quarterback in a huddle calls plays. The play comes in from the sidelines. The quarterback just relays the message. The quarterback needs only seconds to get the message across. You can do the same in about as much time. Be sure the players understand: You're bringing the owner's instructions in from the sidelines. You met with the owner yesterday. Today you're relaying those instructions.

5. *It's OK to order materials if the supplier understands who is responsible for payment.* On the Nenninger job, Joe Marcario gave Phil and Matt authority to order materials from Thurber. Presumably, Phil and Matt worked for Nu Frame Contracting, the framer. That would have been fine if Thurber had understood that Nenninger was responsible for all deliveries to the Nenninger job, no matter who placed the order.

To eliminate any chance of confusion, draft a letter for the owner to sign. Explain to key suppliers who is authorized to order materials for the job. Make it clear that the owner is responsible for paying the bill. Send each letter by certified mail. Get a receipt. Put a copy in your legal file. We'll have lots more to say on this subject in later chapters.

6. *Get started on the right track and stay on that track.* When there's a problem, you may be tempted to take charge and start giving instructions. CM contractors are consultants. They can relay instructions from an owner. But they don't *speak for* an owner. That would be an agency CM contract, as explained in the sidebar later in this chapter.

In a nutshell, your job description is very simple. You advise the owner, get a decision, record that decision, transmit that decision, monitor the process and keep the owner fully informed. To do that, you'll need the best communication possible with the owner. Stuff happens. We'll have more to say on this topic in later chapters.

7. *Don't accept cash payment for construction services.* Instead, have the owner deposit cash at a bank and write a check. That leaves no doubt about payments received and accepted and the date of acceptance. If there's a dispute, proof of payment can be as important as a written contract. Good records win cases.

If you're working for cash as a CM contractor, something isn't right. Use cash to buy lunch. For everything else, leave a paper trail. A CM contractor doesn't need to write checks for anything. Don't even bring your checkbook to a construction site! I'm OK with an owner asking you to pick up some tools or materials at the lumber yard. But there's a bigger story lurking somewhere if an owner wants to pump a couple of hundred grand in cash through a construction project. An audit could leave you with a severe tax headache.

> **Worker's Compensation Insurance for CM Contractors**
>
> CM contractors are self-employed consultants. Worker's comp insurance is insanely costly for most construction trades. Your worker's comp insurance cost should be like that of an architect, not a roofer. An architect's primary risk of an industrial accident is falling off a drafting stool. Your chair is even lower. Your worker's comp premiums will be too.

8. *Worker's compensation insurance is required by law in all states.* Many states also require that licensed contractors carry liability insurance. Premiums for both liability insurance and worker's comp insurance are based on payroll. The higher the payroll, the higher the premium. Worker's comp insurance and liability insurance protect the owner. A good CM contract will require that the CM contractor verify coverage of trade contractors. The owner will decide if coverage is adequate. But it's poor practice for a CM contractor to recommend an agreement with any trade contractor who isn't insured.

9. *Reflect for a moment on Marcario's testimony:* "The plan was for the two to sit down after construction and discuss what Nenninger would owe Marcario for his services." I like that approach — but only if you're doing charity work on an orphanage. For every other job, get a clear agreement on the fee before work begins.

If your agreement with an owner is to settle up at the end, we've got some bad news. You've already settled up. The owner won! A CM contractor has three tasks: avoid risk, hold your leverage, plan to get paid. Everything else is a walk in the park. All three of these issues should be settled by contract on Day Zero. If you do it on Day 1, the owner's already one day's pay ahead of you and gaining leverage. Six months of that procedure and the owner will own the house and you too!

> **Agency CM Contracts**
>
> We wouldn't be candid if we didn't add that there's another type of construction management contract that gives a contractor authority to speak *for* the owner. That's generally called an *agency CM* contract. The property owner gives a hired consultant authority to make decisions for the owner and incur charges for the owner. From a legal perspective, the owner and the hired consultant are essentially the same entity. Agency CM contracts are commonly used on public works projects done for municipal governments. Agency CM contracting is a whole other thing and isn't the subject of this book or what we're recommending.

Finally

We can't resist making a general observation about the facts presented in the case of Thurber Lumber v. Marcario. Nenninger, Marcario and Thurber were all experienced in business. Marcario and Thurber had been doing business for over 20 years. Nenninger was a real estate broker and was working as a mortgage broker at the time. He must have understood the value of a written contract. Why was there no paper trail? The court's opinion doesn't provide an answer. We're not going to speculate about the motivation of the parties in this case. But we've seen other jobs:

- Where the subs don't have insurance and aren't licensed. A formula like that can make you a casualty the day you decide to turn legitimate. For me, no insurance — no work. Keep your life simple.

- Where tradespeople claim to work as independent contractors without tax withholding. That's likely to earn you an appointment in tax court and hearings with the labor relations board and the contractor's license board.

- Where a contractor buys in the name of another contractor to deceive a vendor. This usually means the first contractor is hiding from creditors.

- Where an unlicensed contractor claims to be working for a licensed contractor. The licensed contractor may not even know about the claim. No legitimate contractor needs to take this risk.

- Where the owner wants to pay in cash. It's probably untaxed earnings, maybe drug money. Don't get involved.

- Where the owner doesn't want to take out a building permit. If you get caught, the owner will develop instant amnesia. To get work after that, you'll have to skip to another state.

- Where the owner suggests evading compliance with the building code or local regulations. This always puts someone at physical or economic risk. Don't be that someone.

To preserve deniability on that job, it's likely that very little gets committed to paper. No contract. No receipts. No records. It never happened.

If it never happened, you shouldn't be part of it. That's not paper contracting. In our opinion, it's a foolish way to do business. A job like that can become a stain on the reputation of any construction professional. We doubt any construction company working under the conditions just listed, will grow and prosper for very long. Our experience is that the most reputable, most profitable construction companies wouldn't touch a job with those characteristics. Working as a CM contractor, you already have good immunity from risk. There's no advantage to working on jobs that never happened.

Instead, build your reputation as a competent, effective professional. Construction in your area is a small community. Bad news gets around fast. On public projects, the owner has to check credentials and get references before contracting with anyone. I know. I've been the one who makes those calls. Protect your reputation the same as you protect the PIN to your credit card.

Our Advice

Deal with owners and contractors who aren't trying to deceive anyone, least of all a state, local or national government. In the case of Thurber v. Marcario, Thurber played by the rules and got a judgment for the full amount claimed. Expect the same result when you play by the same rules.

Chapter 3

Managing Pre-Construction Services

MANY CM CONTRACTORS HAVE discovered the advantage of getting involved in a project as early as possible, ideally before design has started. In fact, that's one major advantage of CM contracting. Get in early and you can influence design well before work begins. In public works construction, this is usually called *early contractor involvement*, or ECI. The CM contractor mentioned in Chapter 1 would call this good business. As long as design work goes as planned, it's almost certain that an ECI contractor will participate in construction — and often at a negotiated price.

We're not suggesting that CM contractors get into the design business, though many successful construction companies have a regular working arrangement with an architectural and engineering firm. We're suggesting that there's a role for a CM contractor from the day a project is conceived until final project closeout. If you've never participated in pre-construction services on a construction project, this chapter may be an eye-opener.

A Contractor's Perspective

Why would any CM contractor want to get involved in a project before the plans are drawn? Of course, the most obvious reason is that there's a fee to be earned. Maybe more important, you bring a unique perspective to the pre-construction process. Most owners have no trouble recognizing a design they like. Far fewer are willing or able to coordinate the design services required to develop a workable set of plans and specs. Very few owners will recognize design elements that complicate or delay completion, or that increase costs without adding value. And nearly everyone will agree that a contractor should review a set of plans before design work is considered complete.

The Corps of Engineers uses ECI on many of their larger projects. Levee improvement projects in the New Orleans area are one example. Levee design is done by an engineering firm. But the Corps retains an ECI contractor to work with the engineers, ensuring that the project can be built as designed. The ECI contractor is encouraged to offer suggestions that help control costs or shorten project duration. When design is complete, the Corps often exercises a contract option to have the ECI contractor manage construction at a negotiated price. Even if the project is released for competitive bid, the ECI contractor's grasp of the project guarantees a significant advantage over the competition.

What's true for levee projects in the Gulf Coast is also true for nearly any residential or commercial project — from remodeling a kitchen to developing a shopping center. A CM contractor who participates in pre-construction is performing a useful service, earns a fee for the work, and is guaranteed a seat at the table when the construction team is selected.

Adding Value During the Design Phase

You have a golden opportunity to set a project on the right path during project development:

> ➤ for an owner who doesn't understand the construction process,
> ➤ for an architect whose design may exceed budget limits, and
> ➤ for a contractor whose primary interest is getting a piece of the action.

You either reconcile these three positions right up front, or become a victim before the first spade of earth is turned.

Every architect and every contractor assumes the owner is in control. Nearly all owners are very successful at something, but it's seldom construction. When it comes to running a construction project, most are rank amateurs. Few have any real construction experience. That leaves a vacuum where bad decisions thrive and multiply. Want confirmation? Just watch as designs are approved by the owner. You know the type: grand staircases, round rooms, hand-built heavy timber roof trusses, standing seam metal roofs with copper gutters and downspouts. The owner doesn't appreciate the consequences of approving what the architect proposes. An owner who relies on an architect's cost estimate or budget projections is in for a major disappointment.

Can't a contractor help an owner make good decisions? Some contractors do. But most aren't hired until the plans are done. That's way too late to fix cost problems. A contractor brought in early seldom has real influence over design. More often, the owner and the owner's spouse want what they want and the architect intends to please at any cost. A contractor who complains about cost gets replaced. The owner's hope is that cost problems will disappear when plans go out to bid. That rarely happens. The owner gets lost in the dazzle of design. Advice the contractor gives in public about buildability may be refuted by the architect in private. When nobody is in control, risk multiplies.

Neither the architect nor the contractor wants to take control of the pre-construction. It's not in their economic interests. Higher costs equate to higher fees. And, architects and contractors aren't paid to manage pre-construction. Their role is to design it and get it built. An amateur owner is likely to get an expensive education.

A CM contractor is in a good position to prevent this train wreck. The architect doesn't feel constrained by budget. The contractor, if there is one, just wants the work. That can leave you as sole promoter of a successful project, the only person in the room who really knows how to run a construction program ... not just a project.

To be an effective guardian of the project, set guidelines for all the players, including the owner. Find the most direct and efficient route from concept to completion. Set limits: goals, budgets and schedules. Rein in design before expectations exceed resources. Explain the advantage of simplicity. Start with a flat-roofed, single-story square structure that looks good — not a three-story round building intended to impress from the interstate. Get the contractor hired the same day as the architect. Encourage collaboration. But everyone goes through you; not the owner. Have the contractor price design work weekly. Be sure the architect revises plans based on the contractor's estimate weekly. Everything should fit within the budget and schedule which you control.

Being a CM contractor isn't about managing a project. It's about managing a *process*. That requires control. When done correctly, you control the players without the players realizing that you're in charge. This isn't an ego trip. You're conducting a construction symphony.

Of course, you're the owner's advocate. We'll talk more about that later. But you can also be an effective advocate for the general contractor and the designer. Eliminate problems a contractor or a designer may have with the project. Make the designer's and the contractor's task as simple and efficient as possible. Develop faith in your ability. Impress others with how smoothly you can make the process evolve. Solve problems before others even recognize the problems exist. Job costs will drop when work goes smoothly. Acrimony leads to higher costs. Recognize that managing any construction project is 50 percent science and 50 percent art. The remainder is design and construction.

The Pre-Construction Sequence

Managing pre-construction can include at least six distinct services. Not all of these will be required on every job.

1. Develop an understanding of the project as conceived by the owner.
2. Establish project guidelines, goals, budgets, designs, materials, schedules.
3. Recommend design professionals best able to prepare plans for the project.
4. Work with design personnel to ensure that project plans meet the owner's needs.
5. Recommend building materials, systems and details which allow completion within the time and budget set by the owner.
6. Ensure that the project as designed will comply with applicable codes and regulations.

Finding Prospects

Every construction management project begins with an initial contact. That doesn't happen if potential clients know nothing about your company. Everyone in a service business faces that issue. You need to find new prospects. This isn't a book about selling construction services — but we can't resist making a few observations.

First, simply responding to published invitations to bid will work for some CM contractors, especially those in competition for public works projects. But with competitive bidding come several disadvantages. Lower profit margins is one. The law in most states requires that public works projects be administered under contract documents drafted by the public agency. That leaves little room for discretion. Don't expect to build a successful CM contracting practice simply by responding to invitations to bid. Finding new clients requires an investment of time and effort.

Second, most advertising doesn't work. Experienced CM contractors will agree that TV, billboards, handbills, trade shows, direct mail, classified advertising and similar promotions are a waste of money — and may be counter-productive. Setting up a website might be an exception. Most potential customers will expect that you have both a website and a phone number. A company that doesn't have a website isn't really in business, at least in the eyes of some. You might experiment with pay-per-click promotions in Google *AdWords* or Microsoft *adCenter*. The cost is modest and promotions can be limited to the community you serve. But only rarely will any of these promotions yield a client that turns into a good contact.

What's the best way for CM contractors to promote their service? That's easy. The modern term is *networking* — making yourself known among those in your community likely to need a CM contractor. The more traditional term for networking is developing a *reputation*. Doctors, lawyers, dentists, accountants and architects do this by instinct. The best network you'll ever have is a list of satisfied clients — owners of projects you've completed successfully. Networking is especially important if you're eager to get involved early in project development. You want to be on-site before ink hits paper.

One word of caution: Don't promote a class of work you're not licensed to perform. For example, don't mention engineering or pool construction if you're a licensed general contractor. Where a license is required, state license boards investigate contractors who advertise for work they're not licensed to handle.

The Initial Contact: What's Required?

We're going to assume that an owner has contacted you because of your reputation for professional work. When you get that inquiry, consider it a vote of confidence. Someone has recommended your services. That wouldn't happen if you had a reputation for substandard work. What you do with this opportunity depends on how well you respond.

Every job begins with understanding what the owner needs and is willing to pay for. That requires careful listening. Your prospect has a problem: a site that needs a building or a building that needs improvement. Listen as your prospect explains that problem — exactly what's needed. Be especially alert to likes and dislikes. Many prospective clients will be more emphatic about what he or she *doesn't* like than about what's really needed. Make notes on any expressed dislikes. If you offer a solution that ignores a key dislike, you're likely to get nowhere.

Qualifying Your Prospect

When your prospect has explained what's needed, begin qualifying your potential client. Is this work you want or work you want to skip? Not every job is a moneymaker. You'll hear builders and estimators claim they want (or need) only 10 percent of the work that comes their way. The challenge is to select the right 10 percent.

Many projects will never be built. Don't waste time conferring on projects that aren't going to happen. I call these prayer meetings. You go and pray that something good is going to happen. It's a prayer meeting when:

> ➤ The owner is undercapitalized or isn't a good prospect for potential lenders.

- Code or zoning restrictions make the work impractical.
- The owner isn't being realistic about the cost or what can be built.
- The perceived need is based on assumptions that seem tenuous or transitory.
- The owner has been turned down by several builders or lenders.
- Your prospect may not have authority to contract for the project as conceived.

Within the first 10 minutes, your contact is likely to start asking questions, such as, "What's the best way to do this?" or "What do you think about … ?" or "What would it cost to … ?" or "Do you have a brochure about your company?" or "Can you supply a list of references?" Respond to the question, of course. But also consider this an opening to begin asking questions yourself. The most efficient way to do this would be to have your prospect fill out a detailed questionnaire. But asking for that's probably not a good idea. Instead, work questions like these into the conversation:

- *"Have you talked to anyone about financing?"* Obviously, financing is a key question. Every owner wants to improve their property. Not every owner can qualify for the financing needed to carry a project.
- *"Do you have a budget in mind?"* This is another key question, the beginning of price negotiations.
- *"When would you like see this job finished?"* Identify unrealistic expectations as soon as possible.
- *"Have you talked to any other builder [architect, engineer, or consultant] about this job?"* If so, ask, "What did they say?"
- *"Have you considered … ?"* Try to identify zoning problems, potential issues with neighbors, design review committees, setback requirements or anything else that could halt the project.

If the initial contact isn't on the proposed construction-site, make an appointment to meet on-site. There's no substitute for walking the ground where work will be done. You need to know about potential problems with access, unsuitable soil, neighbors, location of utility lines, drainage and general site condition.

Who Gets the Work?

CM contractors should consider this engraved in bronze: *Owners sign contracts with those they like and trust.* That's simple human nature. Do lunch. Offer to buy dinner. Fit in. Make an owner's business life your business life. But this social quotient is like a coin that can fall either heads or tails. The lowest bidder may not get the work. That's good news, assuming you're not planning to be low bidder. The bad news is that you may submit a very

competitive, highly innovative offer that's rejected on grounds not clear to you, or that the owner won't explain. Maybe, for example, the job went to the owner's brother-in-law or an old fraternity brother. You'll never know. Owners aren't eager to admit overt preferential treatment. But outcomes like these are a risk in every competitive selling situation. Unfortunately, there's no foolproof way to guard against risks that defy conventional logic. Private owners, unlike most public agencies, aren't required to award work to the lowest responsible bidder. At least occasionally, even your best efforts aren't going to be rewarded.

Experienced contractors usually agree on what it takes to sell construction services in a competitive market:

➢ Be the most thorough, most complete, most diligent competitor. If you ask owners, especially private owners, why they selected a particular contractor, the most common response will be, "They gave me a good proposal." In the eyes of an owner, a contractor who doesn't respond completely or as expected isn't likely to complete the job as expected.

➢ Be friendly and likable — someone the owner would consider a good contact. No one wants to disappoint a friend. Be a knowledgeable professional resource.

➢ Provide something unique — an insight or option the owner didn't consider. You've probably won the job if an owner likes one of your suggestions well enough to request the same feature from other contractors.

➢ Respond 100 percent to every concern. The essence of salesmanship is eliminating objections. If necessary, ask the question, "What do I have to do to get this job?"

Don't Be Afraid to Decline the Job

Ask yourself these questions before agreeing to submit a proposal.

Is this the type of work I really want? It's poor practice to take on work that you consider marginal. Even worse than not having enough work is getting mired in a job likely to end in a loss, an emotional dispute, or a court case. If the job isn't big enough to warrant your best effort, look for other opportunities. You'll hear jobs like this referred to as *church jobs*. Every time you step on the jobsite, you're making a donation.

Is this the type of work I've handled successfully in the past? The best guarantee of future success is a good record on similar projects already completed. Taking on a new type of work can stretch company resources. Your time and technical skills are limited. It's better to stay within your comfort zone.

Has the owner come to you as the result of a referral? Or did you have to make a cold call to find this work? Your proposal always carries more weight when a third party suggested the initial contact. The more influential, the more respected that third party, the better your prospect of getting the job.

Do you have a pre-existing relationship with the owner? People you know socially or professionally can be the best business prospects.

What's the risk? Competing for a large, complex project can require a major commitment of time and effort. If competition for this job will require weeks or months of work with no guarantee of success, you may not be prepared to make the commitment required.

Is there something about this job that gives you a clear advantage over the competition? That advantage could be similar jobs completed in the past, either for this owner or for others. It's a plus if you have a good working relationship with an engineering firm, an architectural firm, specialty contractors, or vendors who specialize in this type of project.

What does your instinct say? As humans, we make both instinctive (emotional) decisions and decisions based on reason (logic). Instinctive decisions take only seconds. In a heartbeat, you know whether you want this work. When instinct confirms logic, you'll feel comfortable making a decision. But logic and instinct can pull in opposite directions. When in doubt, rely on instinct. Here's why. Instinct is usually a better judge of people and character. The potential for profit or loss varies with the character of those involved.

Sell Yourself

Assuming you've decided to compete for the job, it's time to blow your own horn. This shouldn't be simple boasting or high-pressure selling. The more subtle your approach, the more likely you'll be understood. Tone, attitude and demeanor say a lot about competence. To develop trust, offer information the owner may not have considered on topics the owner probably doesn't understand, such as building codes, zoning, permits, choice of materials, and potential problems. Make it apparent that you're a good resource on all these issues. Describe the jobs you've completed and your familiarity with trade contractors active in the community. Mention local planning and building department officials by name. To increase faith in your judgment and value to the project, identify resources or authorities you could use on this project.

Don't be afraid to admit you don't know the answer to some question. No one knows everything about construction. Instead, be frank. Admit that you don't know. Then explain that you'll have an answer tomorrow or the next day. Promise to call back or send a response by email. The advantage should be obvious. Every sale of construction services begins with a dialog between a contractor and the owner. The best jobs will require multiple contacts over several days or weeks. The more often you exchange information with the owner, the more likely you are to get the work. Multiple contacts will reassure an owner: You're available and ready to help when needed.

Keep a marketing notebook with lists of leads and contracts. If you've got a hot lead, set up a notebook just for that lead. Use dividers to separate various points that develop with that lead. Note questions asked and answered, information requested and supplied, and the dates when each happened.

The Cost Question

What's your response when the owner asks about the cost? This is where construction is different from nearly every other industry. Everything you buy at a store or on the Web has the price set after manufacturing is completed — and costs are well known. There's a price tag on nearly everything we buy. Even professional services come with a price tag, such as $6 an hour for a babysitter or $300 an hour for an attorney.

Construction is different. With construction, the price is set before production begins. That's a disadvantage in a commercial environment where cost is an important part of every decision to buy. Builders have to attach a cost before production starts. That increases the risk inherent in construction contracting.

If you quote a price that's higher than expected, the job may not happen. If you quote a price that's too low, your casual comment could end up being quoted in a legal brief filed by opposing counsel. If you refuse to quote any price, you'll be considered devious or uncooperative. So what should you say when an owner asks about cost?

No single answer works in all cases. But one point should be clear: You can't quote an exact price at this stage. Even if you have a ballpark figure in mind, consider keeping that number confidential. Instead, consider one of the following responses.

> ➢ "That depends a lot on what you decide. It's a little too early to nail down a price. But I'm sure we can live within your budget. What figure do you have in mind?"
>
> ➢ "I've seen jobs like this go for between $X and $Y. Of course, the cost could be less or more, depending on choices you make later. A lot depends on finish materials and when you want to get started. When I know more about the job, I'll give you a written estimate."
>
> ➢ "I don't want to quote a number off the top of my head. But I have some good references back at the office. I'll work up some numbers based on those figures and get back to you tomorrow with typical square foot costs."

Of course, any answer to the price question depends on your role. If you want to be the contractor, every question about cost carries risk. You may have to complete the project at that price. If you plan to be the CM contractor, break any answer to the cost question into units. For example, how much does a new house cost? "Well, the cost should be under $200 a square foot, plus design fees, the lot, and financing. My CM fee for running the project will be $100 an hour plus expenses. Generally that works out to about three percent of the total project cost."

Planting a Seed

This book is about CM contracting. By now, you're very familiar with the concept. But the owner you're talking with may never have heard the term. If you've decided that the job is a good candidate for a CM contract, now is the time to broach the subject. We don't recommend that you offer a CM contract on a take-it-or-leave-it basis. But there's no harm in suggesting a CM contract or offering to bid the job two ways. As you'll see later, that's easy.

> *"Describe how a CM contract works*
> *— multiple prime contracts."*

Raise the subject by suggesting that a CM contract could be the right choice for this job. Describe how a CM contract works — multiple prime contracts. "I can save you some money on this job. I'll act as your consultant at a price my competition can't touch. There's zero profit for me in this job. Just pay me for my time. I'll give your job the same attention I give to every project I take on, but with no markup. It's hard to beat a deal like that. I'll recommend the design team and work with that team until you agree that their plan is perfect. I'll make sure the plan is done right and complies with the code."

Don't leave the site before getting the information needed to draft the contract:

- name and address of the property owner
- construction-site — either street address or legal description (for lien purposes)
- phone numbers — business, cell and fax
- email address
- a good concept of what the job requires — including rough dimensions and square footage
- potential access issues, the availability of water, electric, sanitary facilities, etc.
- permits required to process the plans
- utility companies that will be involved

The Second Meeting: Make an Offer

When you return to the site, have at least a conceptual plan of action likely to impress the owner. Come prepared with a draft CM contract for pre-construction services ready for signature. Your prospect probably hasn't paid anything for this work. So what you bring to this second meeting doesn't have to be more than a rough diagram of the overall process. But the CM contract should be ready for signature. More about that later.

Of course, every job is different. So every sales pitch has to be different. But every sales presentation made by a CM contractor will have common elements. We're going to offer an example, an actual transcript of a successful proposal made for a pool contract in the Sacramento area of northern California. We'll call the CM contractor *Jack the Pool Builder*. For reasons that will be obvious later, this isn't the CM contractor's real name.

Jack the Pool Builder

Jack agreed to meet with the homeowners on-site (a new home under construction) a few days after an initial phone contact. Jack arrived at the site with his sales kit: pictures of pools, color samples and design sketches. He was ready with recommendations on square footage, location of the pool's "swim out," a spa, a concept of how the pool should be oriented on the site and a location for the pool equipment enclosure. The owners agreed that Jack's plan met their needs and particularly liked some features. Jack explained that changes to the initial design could be made later. Jack told the homeowners he would "finish all the designing, all the inspecting, get it all engineered ... I'm going to quote out the contract for each stage of the process to the same guys I use for every pool. I'm going to get you all the quotes — everything."

Jack offered to provide the homeowners with a schedule for each stage of construction and a list of contractors who would do the work. He would also provide release of lien forms when each contractor completed work and would explain how to complete those forms.

Jack:

> The other packet I plan to give you is going to have like a blueprint, kind of a step-by-step of who comes in at what time and how long each contractor should take. I'll help you schedule those guys. I'm going to put the whole package together, do all the grunt work. Each contractor will have your plans. They'll know what to do, where to start, all the measurements, everything. All you say is OK, OK. Start with the excavation guy. He'll come in, dig a hole and finish up. When I tell you he's done, you write him a check. I'll have him sign a release saying he's been paid. Then the next guy comes in. I'll make out a schedule of contractors — about three at a time.

Homeowner:

> *What's your role in this job?*

Jack:

> *You're paying me for getting all of the designs, all the specs and all the engineering done, getting all your quotes from the contractors and basically putting you through the whole process, putting all this together. I quote it out to everyone. I'm getting the quotes for you. When I bring the packet to you, you'll have everybody's exact quote with their license numbers. I'll come back and give you that packet along with two or three quotes from these contractors ... so you'll have their quotes. I have a relationship with these guys, I mean if you cold-call one of these contractors, they might charge $7,000. When you go through me, you'll get the same job for $6,000. You get the benefit of my relationship with these contractors. But it's even better. I'm working for you. I'll be the guy on the job looking out for you, protecting you. That's what I do.*
>
> *I have 16 pools right now on the go. Our goal is to do 42 pools this year, and we'll probably get 60.*
>
> *The guys I use are the best of the best. Other guys can do this stuff and other guys want to dig and do stuff for me. They would be cheap. I could get pools in cheaper. But they're not as good. These guys are my guys — been doing this for 15 or 20 years. The last thing they want to hear is somebody say, "I'm calling the State Board."*

Jack offered the homeowners a construction management contract for pre-construction services. We'll have more to say about this contract a little later. After asking for a few changes in Jack's preliminary plan, the homeowners signed Jack's CM contract for pre-construction services and gave him a check to cover the first payment identified in the contract. If the homeowners approved Jack's design and agreed to go ahead with the pool job, Jack would credit his fee for pre-construction services against the cost of construction.

Jack said he'd be back later in the month with working plans, a bid price, and a draft agreement for the project.

The Third Meeting: Get a Commitment

About two weeks later, Jack called to say he had the plans and specs ready and wanted to schedule a meeting to present the proposal. Jack arrived on-site with both the plans and a draft CM contract. The homeowners went over the plan carefully, made a few suggestions and then asked about the cost. This time, Jack was ready with numbers.

Jack:

> *You'll pay on this pool here with everything, all the equipment, all the decking, everything, $47,300. That's the top price it's going to cost you, with all of your equipment.*

Homeowner:

But do we have a quote, do we have any idea how much it's going to cost?

Jack:

$47,300 including all equipment, everything, from start to finish. The only thing you'll pay for beyond that is for your permit ... My fee is $5,000, which is included in the $47,300. What you've paid me so far for the plan and engineering is a credit against that $47,300. Pay the rest of my fee as the work gets done. It's all in the contract — which I have right here. The first payment is due when you say, "We're ready to go ahead." Make the last payment when you agree that the job is done just the way you want it. Between those two checks, I'm your support guy. You'll be calling, 'cause in the first two or three stages you'll be asking, "Now who? Now what?" It goes that way with everyone. And by the time the coping goes in, people are just flying with it.

Homeowner:

And, if we have a problem, who do we call? Do we call you?

Jack:

You'll call me. You need to be able to call someone. Yeah. You call me. If there's a problem or something you don't like, just call me. These guys jump when I tell 'em to jump. If it's like not showing up or doing something wrong? Well, that doesn't happen.

Homeowner:

Do you come out all through the process?

Jack:

Yeah, I come out every couple of days. I'll come out and check up on things and make sure things are flowing. Remember, I have just one job, to protect you — to make sure the job gets done right.

Homeowner:

I know I'm to line them up. But you'll be out here to check 'em out a little bit.

Jack:

You've got seven stages. A couple of times at every stage I'll be here. Before we go to pre-gunite, which requires an inspection, pretty easy, I'll be here. If the rebar guy did his job, I'll call for inspection. The inspector shows up the next morning. So before pre-gunite, I'll be here. Before they pour the deck, I'll be here.

Homeowner:

> You'll check that out?

Jack:

> I'll come and check it out.

Homeowner:

> And, you call us and say, "Hey we've got to fix this or whatever?

Jack:

> I'll do that. My job is to make sure you know what's happening. That's in our contract. Look right here. The agreement says, "Keep Owner Informed of Progress." I wouldn't sign this agreement if I didn't plan to live by it. Here's what my fee covers. It's right here in the contract.

Jack's sales pitch was letter perfect — very successful. But unfortunately, it wasn't for Jack. He wasn't licensed for pool installation. What he got for his efforts was a $2,000 fine from the State Board — for contracting without a license. Jack made this presentation to two undercover investigators from the state license board posing as homeowners.

We hope the experience motivated Jack to apply for a pool contractor's license. He's very good at what he does. He just needs a license to do it.

Jack's Formula for Success

Notice a few points about Jack's approach to CM contracting:

It's a package deal. Some clients won't be eager to get into multi-prime contracting. They would prefer to have the entire job done by one general contractor. Jack's "packet" addresses that issue. It's a roadmap for the project — who to call and when, what to expect next, reassurance that Jack will be on-site and in a support role when needed.

Jack as the owner's advocate. This is an important point that distinguishes general contractors from CM contractors. Owners and general contractors are natural adversaries. Anything demanded by an owner cuts into a general contractor's profit margin. CM contractors are the owner's advocate. It's a point not always understood by owners. Jack made the point twice during his presentations.

Everyone loves saving money. Jack had a powerful argument. "There's zero profit for me in this job. Just pay me for my time." Jack's competition won't be eager to discuss their profit margin. Jack put the issue right on the table. If Jack is the only contractor to offer a CM contract, he'll probably get the work.

Managing Pre-Construction Services 51

Make a smooth transition to the contract. Once the owners are sold, focus should shift to the contract. It's simply good practice to explain important points in the agreement. There will be questions. Better to explain key points before questions come up. We don't recommend asking for a signature immediately. Better to leave the contract with an owner and suggest that you'll call back tomorrow to answer questions. And remember, if work is done on a primary residence, the owner has three days to cancel the deal. More on that in Chapter 6, *Reviewing Trade Contracts*.

"Gentlemen, we've decided that it's time to apply for a contracting license."

It's a three-step process. Notice that Jack made three presentations. The first was a phone call and ended with an appointment to meet the owners on-site. At the second meeting, Jack had samples, examples, and a CM contract for pre-construction services. Most of what went on was careful listening — by Jack. At the end of the second meeting, Jack asked for a commitment and got one — a signature on the CM contract for pre-construction services. The third and final meeting should have resulted in a go-ahead to start the job. Jack had the plans ready. He had quotes from trade contractors for each part of the job. He had agreements with each of the principal trade contractors ready to sign. He had a job schedule. The only thing he didn't have was a pool-contracting license.

Jack's First CM Contract — Pre-Construction

We haven't seen the agreements Jack used. But from the transcript you've just read, we can surmise that the cost quoted in the first agreement covered:

➤ *Pre-construction services.* That was primarily the design and engineering required for the job — work done by professionals on a fee basis. Pre-construction services can also include an estimate of project cost and a schedule. These topics will be covered in the remainder of this chapter.

➤ *A review of the plans and specifications.* Somebody has to be sure contractors can build what's been planned. As a CM contractor, that's Jack's job — getting the kinks out, identifying errors and code problems early in the process, ensuring that the plans meet the owner's expectations. The result of this review will be a set of plans and specs that require very few changes once work starts. We'll cover plan review in Chapter 4.

➤ *Bid packages and an evaluation of the bids.* Jack returned to the site for the third meeting with a firm bid on the job, $47,300 (including Jack's fee). Jack had firm bids from the principal contractors. Those contractors had reviewed Jack's plans and submitted bids. From the transcript, it seems Jack had at least two bids for some of the work. That's a nice touch. Jack and the owners needed to discuss those bids. Jack would make a recommendation and the owners could decide which bid to accept. That's the way CM contracting is supposed to work. Solicitation of bids and evaluation of bid packages is the subject of Chapter 5. On a large project, the collection of bid packages may fill a three-ring binder. You may have as many as five bids for each part of the project: earthwork, concrete, framing, wiring, plumbing, mechanical, windows and roofing. On a project for a public agency, a summary sheet will show which bidders are local and which are from outside the area — usually confirming that local vendors get the benefit from local bond sales. That's good politics.

> *Contracts with principal trades.* Jack doesn't say so, but it's fair to assume that his "blueprint packet" included contracts with at least several of the trade contractors. In some states, law requires a written contract if the value of residential work exceeds a certain amount. Almost certainly, Jack drew up these contracts. Without doubt, Jack was ready to make a recommendation to the owners on each contract. That's part of the CM contractor's job. If you understand what the project requires, Chapter 6 will explain how to prepare these contracts. Jack may have had a dozen contracts in the "blueprint packet." But all would have been based on the same model agreement stored on Jack's computer. Using dozens of different contracts on the same job would be a real nightmare. For my part, if a contractor doesn't want to use my contract, I don't use him. You have plenty to do without crawling through a pile of dense contract language every time there's a dispute. About half a dozen standard contracts could get nearly anything built. For design work, I use a master contract that defines responsibilities but doesn't get specific on the work, the price or the schedule. As the owner releases design work for completion, I issue an addendum to the design contract defining the work, money and schedule.

Jack's Second CM Contract — Construction

From the transcript, we can guess what was in Jack's second contract, presented at the third meeting, just before the undercover agents flashed their State Board credentials.

> *Managing construction.* Jack emphasized this in his sales pitch. He would be on-site several times during each phase of construction, checking on the work, ensuring that the trades were making good progress, anticipating problems. If you've worked on construction-sites for more than a few years, most of

this will come as second nature to you. In Chapter 7, I'm going to suggest that you be on-site daily. Find and head off problems before anyone knows there is a problem. On larger projects, I'm on-site twice a day for at least two hours, mid-morning and late afternoon. I know something is going wrong long before the contractor, architect or owner. When they hear about a problem, it's usually a problem I solved two days ago. They're just playing catch-up.

➤ *Keep the owner informed of progress.* It's simple human nature to wonder about what you don't understand. Owners want to be kept informed. Jack was quick to point out reassuring words in the contract. Chapter 8 suggests good ways to keep information flowing smoothly. For example, I try to get my face in front of the owner's face three times a week — and daily if I can find enough excuses. No competitor is going to steal your job while you're in conference with the owner. On a big job, I offer to take board members on a tour of the site as often as they're willing. Board members are owners too.

➤ *Evaluate payment requests.* No owner wants to get into a dispute with a contractor about payment. On larger public projects, an architect may be responsible for approving requests for payment. That may not be the best solution on all jobs. Architects and contractors are potential adversaries — one claiming that the plans and specs are clear and complete; the other claiming the plans are ambiguous and that extra work is required. A CM contractor makes a better impartial third party on most construction projects. Chapter 9 will describe a CM contractor's role in approving payment requests. Ideally, the owner, the architect and the contractor work as a team. But when it comes to paying invoices, that love triangle can turn into a vicious blame game. The owner gets backed into a corner as referee. The standard construction process doesn't inherently promote teamwork. There are too many competing self-interests for the work to progress smoothly at all times — especially when it's time to pay bills. The CM approach gets the owner out of the referee business. That's a great selling point for any CM contractor.

➤ *Communicate with contractors and suppliers.* This is where traditional general contracting and CM contracting differ in theory more than in practice. A CM contractor is a consultant to the owner. In theory, a CM contractor doesn't have authority to give instructions to contractors, and doesn't order materials. In practice, Jack said it best, "These guys jump when I tell 'em to jump." Still, it's important that both trade contractors and suppliers understand: A CM contractor has no authority to either give instructions or place orders. The owner retains that authority. In Chapter 10 I'm going to suggest taking on as much responsibility as the owner will allow. Anything less can be a drag on job progress. I usually extract an agreement from the owner that I act as the owner's alter ego, nearly like the owner's agent. For example, on a big job, I might request authority to make agreements with contractors and designers up to $100,000. Recently, I did a large project with several contractors and architects that were providing pre-construction services spanning several years. Their prices were coming in sky high. The further I asked them to quote prices into the future, the bigger the numbers grew. So I stopped asking for quotes. Instead, I put them on a money-rationing system. The owner gave me authority to commit for up to $100,000 at a time. When that money ran out, the designer or contractor had to come to me for more cash. After three or four return trips to my money tree, they were ready to break ground. In the end, I saved the owner $500,000 on pre-construction services.

> *Assist with change orders.* Every request for a change order has the potential to deteriorate into a construction dispute. The most common change order problem is a dispute about what the plans require. The contractor insists that some feature isn't covered in the plans or specs. The owner (or inspector) insists that the same feature is either implied by the plans or required by the code. This is where a CM contractor earns his or her fee. Chapter 11 has some good rules to follow when dealing with changes — including my rule number one: Settle change orders within the week. It's hard for a contractor to keep arguing about a change order when the check has already cleared the bank.

> *Assist with construction claims.* Every CM contractor has the responsibility to reduce or eliminate disputes. In practice, claims are almost inevitable. In Chapter 12, I'll describe a CM contractor's role in settling disputes. Nearly living on-site will resolve about 90 percent of the most common construction claims. Being on-site, you can collect information and answer questions immediately. When you sense a claim is developing, offer to buy lunch. Settle the claim over a cup of coffee. Constructions types hate paperwork. No one wants to work until midnight on a construction claim. I'll have lots more to say on this in Chapter 12.

THE SCOPE OF WORK FOR A CM CONTRACT

Every CM contract you see will cover some or all of the following:

> *Design & engineering —* Chapter 3
> *Review of the plans & specs —* Chapter 4
> *Solicit & evaluate bid packages —* Chapter 5
> *Contract with the principal trades —* Chapter 6
> *Manage construction —* Chapter 7
> *Keep owner informed of progress —* Chapter 8
> *Evaluate payment requests —* Chapter 9
> *Communicate with contractors —* Chapter 10
> *Assist with change orders —* Chapter 11
> *Assist with construction claims —* Chapter 12
> *Manage project closeout —* Chapter 13

> *Manage project closeout.* Someone has to walk the project, make notes and draw up a punch list of work not complete. No one is in a better position to do this than the CM contractor, as I'll describe in Chapter 13.

Q & A on CM Contracts

Q: Could Jack have used just one contract?

A: Of course.

But consider the problem. Jack offered to quote a set price for the job — a guaranteed maximum price. Committing to a firm price for the work before plans are drawn would be like writing a blank check — almost certain to end in a dispute (or a financial disaster). As we discussed in Chapter 1, CM at risk *(CM@R)* contracting is a perfectly legitimate way to do business. But CM@R assumes the job is well-defined by the plans and specs. Unless you can eliminate nearly all the cost variables in a job, you should get complete plans and specs before quoting a contract price. Jack's first contract covered the design and engineering. Once the scope of the work was clear, Jack could get firm bids from trade contractors. The second contract included a guaranteed maximum price.

A single contract for CM services would be fine if Jack hadn't offered a guaranteed maximum price for the job. By the end of the second meeting with his clients, Jack knew enough about the job to write a CM contract — so long as there was no guaranteed maximum price. But most property owners want to know the cost before work starts. Six states (CA, IL, MA, NV, PA and TN) require that contracts for home improvement work (including pools) show the total cost in dollars and cents. If you plan to quote a firm price before plans are drawn, count on two-step contracting.

Q: Still, having two contracts seems like overkill.

A: Look closer at the advantages.

The first contract required only a small commitment; hundreds, not thousands of dollars. Many owners will be comfortable making an initial commitment like that, especially when the amount collected can become a credit against the construction cost. Don't underestimate the value of asking for a commitment early in the selling process. An owner who isn't serious won't commit, even for a few hundred dollars. Better to discover this as soon as possible. That's what I call the *smoking gun* method of selling. Pull the trigger. Ask for the money. Either close the deal now or blow the prospect out early and get on with something else that makes a profit.

Using two contracts has another advantage. Design and engineering are professional services. That never comes cheap. It's entirely appropriate to ask for cash up front. Property owners expect to pay for professional design services.

Any time you ask for money up front, expect that a contract is required. Many states (CA, CT, DC, HI, ID, IL, IN, LA, MA, MD, ME, MI, NJ, NY, PA, RI, TN, TX, WA, WI, WV) currently prohibit collecting in advance for home improvement work (including preparation of plans) without a written contract. Federal law requires that you provide the three-day notice of right to cancel (Reg Z notice) if work on the principal residence of an owner could result in a lien on that property. That includes every state because all states grant a mechanics' lien for work done to improve property.

"Remember folks, this is just our first contract."

"Any time you ask for money up front, expect that a contract is required."

Finally, when you're negotiating on the construction-site, 26 states, currently AK, AL, AR, AZ, DC, FL, GA, HI, IN, KY, ME, MI, MO, MS, MT, ND, NH, NY, OK, OR, PA, RI, TX, VT, WI, and WV, consider the transaction a *home solicitation sale* and require another set of written disclosures.

Put simply, you need a written contract for pre-construction services.

Q: Suppose the owner already has a set of plans. What then?

A: Just pick and choose the services you offer – cafeteria style.

The sidebar we looked at earlier lists twelve distinct tasks in a CM contractor's portfolio. What you offer depends on what the owner needs. If you use contract software, like Craftsman's *Construction Contract Writer* program, this is easy. Just click the options you want to include in the scope of work. The rest is automatic.

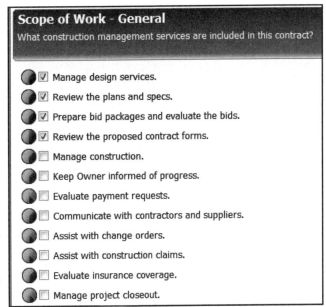

Construction Contract Writer
CM Contract Options

Click to identify the scope of work.

Notice that Jack's first contract included the first four tasks on the Scope of Work. We can assume that his second contract included most or all of the last eight options.

Your Design and Engineering Team

If you offer a contract for pre-construction work, you'll need help with architectural and engineering services. Note that most design services require a license in either architecture or engineering. You can talk *generally* about design and engineering all you want. But only a licensed architect or engineer can prepare a fee proposal and do drawings for structural work.

The architect's and engineer's responsibility is to prepare plans and specs that:

> ➢ meet the needs of the owner,
>
> ➢ qualify for issuance of a building permit, and
>
> ➢ can be followed successfully by the contractor.

We're going to use the term A/E to refer to architecture and engineering. Architecture and engineering are different, of course. Architecture deals with form, appearance, use of space, color, texture and utility. Engineering is concerned with strength, capacity and durability. Your design firm will identify plan features that require engineering calculations to comply with the applicable code. The building department that issues the permit for your job will require the stamp of a licensed engineer certifying that the materials and design selected comply with accepted standards.

If you've never worked with an architect or an engineer, the starting place will be to select an A/E firm that can help with pre-construction planning. On larger projects, especially projects developed at public expense, the CM contractor or public agency will publish a request for proposal (RFP) for A/E services. Competition will be open to any firm that qualifies for bidding. On smaller jobs, you'll have to recruit an A/E firm willing to take on the work.

Selecting the A/E Team

You're probably familiar with the names of architects and engineers active in your community. Nearly every set of plans you've seen includes a name and contact numbers for a designer, architect or engineer. You probably know which A/E firms handle specific types of work. And you probably have some preferences among A/E firms based on plans you've followed. That's a good place to start selecting the A/E team best suited for your project. If you draw a blank, search on the Web for architects, draftsmen or civil engineers in your city. You'll find an extensive list. Narrow that list to about three good prospects.

Like construction companies, most A/E firms are open to meeting with prospective new clients. Selecting the A/E office that's right for your job is like selecting a lawyer or a doctor. Call and ask for an appointment. Explain that you're going to need A/E help with a job. Explain that no plans have been prepared yet and that you expect to act as CM contractor on the job. If you've got a signed contract for pre-construction services, be sure to mention that fact.

Explain that you're going to recommend the design team for this project and want to make a good choice. The owner will sign the contract for design services, will have final approval of the design and will be the client for contract purposes. But as CM contractor, you'll review and approve the plans before they're put out to bid, and will recommend trade contractors who will do the work. Describe the advantages you bring to this relationship: You're familiar with construction and design and will always be available to resolve issues as they come up.

There's no need to mention your client at this point. But it's prudent to ask at the outset if the A/E firm has any reservations about working with you as a CM contractor. Some architects and engineers may not feel comfortable working with a CM contractor. They would rather work directly with an owner. If the firm isn't willing to respect your position as a CM contractor, move on to another office.

Qualifying an A/E Prospect

Ask about work the A/E firm has done in the community — including a list of completed projects. Most A/E firms can provide an attractive color

"Something tells me this architect's going to strain our budget."

portfolio of recent work. If your client needs reassurance about qualifications of the A/E team, this portfolio should settle the issue. Pay special attention to projects most similar to the work at hand. You don't need a hospital design firm for a room addition.

Explain to the A/E firm what your client needs. Explain how you want to handle the work. Be specific about the preliminary drawings you'll need. Ask who will actually prepare the drawings. The A/E representative you're talking with probably handles sales for the firm. If there are more than a few professionals on staff, actual work may be done by someone else. Ask about workload. If your client is in a hurry, don't rely on an A/E firm that can't make a time commitment.

Don't be reluctant to interview small firms. Many very competent A/E firms consist of a single architect or a single engineer with little or no office staff. That tends to reduce overhead. Larger firms will have a more diverse pool of talent on call. But if your job doesn't require multiple disciplines and several experts, a small firm may have everything that's required.

Grounds for Selection

Favor an A/E firm within easy driving distance of the proposed project. A jobsite visit isn't essential for the design of every project. But you want an A/E firm eager to develop a reputation for excellence in your community.

Favor an A/E firm that can point to at least three similar projects completed in the last few years. Ask for references. Call these references. Does that owner have a generally favorable impression of work completed by that A/E firm? If doing a similar project again, would that owner select the same A/E firm? Was the project completed on time and within budget? Was the A/E firm involved in any significant dispute during construction? Ask about changes required after work started. Construction is complex. Changes are nearly inevitable. Some changes aren't the fault of the A/E team — such as changes required by the owner. But if other changes were required, were they due to code requirements (inspection) or due to change orders requested by the contractor? Too many of those may indicate that the plans were ambiguous. Ideally, the cost of changes should be less than 1 percent of the finished cost.

Favor an A/E firm that has worked successfully with CM contractors. But remember that CM contracting is relatively new. Don't exclude a firm from consideration simply because they've never worked with a CM contractor.

Ask for a specific proposal:

> What can the firm do to meet the owner's needs?
> About how long will it take to complete design work?
> What documents will be required to get a building permit issued?

Don't end the discussion without covering fees. There won't be any charge for this first interview. But expect that every contact thereafter will come at a price. After all, A/E firms have nothing to sell but their time. Be sure you understand whether fees will be based on an hourly rate or a percentage of job cost, or some combination of the two.

Most problems that become obvious during construction have their origin in the design phase. Construction is too permanent and too expensive to start over if something doesn't go right. Plan intensively and then execute ruthlessly.

Finally, consider how you and the owner are going to feel about working with this A/E firm. That calls for a subjective evaluation. Still, your response has to be an important consideration in the selection process.

Contracting for A/E Services

Each A/E firm you interview will probably offer a "standard" contract sold by one of the professional architectural or engineering organizations. Beware when anyone refers to a contract as "standard." We believe the term *standard* when applied to contracts is often a deliberate attempt to deceive or short circuit negotiations. The implication is that you don't have to read and can't change a "standard" contract. Neither is true. There's no single standard contract for design services just as there's no single standard contract for construction.

"We believe the term standard *when applied to contracts is often a deliberate attempt to deceive or short circuit negotiations."*

Any printed contract for design services is likely to be biased in favor of the A/E firm. It will also be pretty complete on the scope of work — what has to be delivered, and when. To avoid misunderstandings, read the agreement carefully. Be ready to point out to the owner any sentences or paragraphs that don't apply or that could be crossed out. It's perfectly legal to make handwritten changes to a printed contract. Just make sure your changes can be understood. Then have both parties initial the change. If necessary, write up changes on a separate page and make a note in the printed contract referring to the changes. That's called *incorporation by reference* and is perfectly legitimate. Again, be sure both parties initial the change.

If design work requires a significant investment, more than a few thousand dollars, it's prudent to have work done in stages. Design involves personal preference and is very subjective. If the design process isn't going as planned, the owner needs options — points when the owner can stop design work, pay for services rendered to date, and make other arrangements. That's called *termination at the owner's discretion* and should be part of any contract for design services. Giving the A/E firm the same option to back out at their discretion will help balance the scales. On termination by either party, work completed to date should become the property of the owner once the A/E firm has been paid for that work.

When you've selected the A/E firm and understand their contract, you're ready to make a recommendation to the owner. Like Jack the Pool Builder, you'll need a preliminary design, or at least some sketches, to make a persuasive presentation. Unlike Jack, you may want to make this presentation to the owner at the office of the A/E firm selected.

The Design Process

Once the owner has signed an agreement with the A/E firm, your work can begin in earnest. The first task will be ensuring that the A/E firm understands the owner's concept of what's required. The best way to ensure good coordination is regular communication.

- ➢ Make yourself available to consult with the design team as needed.
- ➢ Schedule regular progress meetings if you feel that will help.
- ➢ Be prepared to take significant decisions to the owner when you're not sure what the owner prefers.
- ➢ Be wary of significant changes in the scope of work that could increase cost or delay completion.
- ➢ Ask for early notice from the A/E firm if design work is falling behind schedule or exceeding budget.
- ➢ Get cost estimates from a contractor, not an architect. That helps protect both the design team and the owner.

Act as an efficient buffer between the owner and the A/E firm. Assume that every change required by an owner will come at a price — both in design cost and construction cost. There should be no surprises. Don't let work begin on any change until both you and the owner have a good understanding of how that change will affect both cost and schedule.

When design work will require more than a week or two, the A/E firm will probably ask for progress payments as work progresses. You have responsibility for approving these invoices and ensuring that amounts billed (1) reflect work completed and (2) comply with terms of the design contract. Keep a spreadsheet with running totals:

- the contract price for all design work,
- (plus) the amount of any approved changes,
- (minus) the amount billed during prior periods,
- (minus) the amount billed in this period,
- (totals) the amount still due on completion of design work.

If you feel it will be helpful, estimate the percentage of design completed to date. Compare your estimate with the percent of the contract price paid to date. These figures should correspond very closely as you approach 100 percent in both categories.

Make this information available to the owner. The easiest way is to include a copy of your spreadsheet as an attachment when forwarding each invoice from the A/E firm.

Design work isn't complete, of course, until:

1. You're satisfied that work can be completed as planned and within budget.
2. Questions from potential bidders have been resolved.
3. The permit has been issued.

We'll cover both your review of the plans and project cost estimates in the next chapter. But we need to mention here that plan review should begin long before design is completed. There shouldn't be any surprises. Any A/E firm should welcome your help in spotting design problems before those problems become expensive design mistakes.

How Architects Work

Design work is done in several distinct phases.

- Phase 1 is usually referred to as *programming*. The project is defined in words: building area, functions, relationships, materials, budgets and schedules. This narrative becomes the scope of the work.

- Phase 2 is the schematic design. The written scope of work is turned into diagrams and drawings. This is the first visual representation of what's going to get built.
- Next comes the preliminary or design development phase. Hard line drawings will show exact areas. Rough specifications may also be part of this phase.
- Working drawings and construction documents follow. These are the plans used by construction crews and will usually include both engineer's drawings and final specifications.
- During the bidding phase, contractors are invited to price out the project. Building permits are usually processed during this phase.
- During construction administration, an architect may monitor construction for compliance with the plans and specs.
- The final phase will be the punch list and project close out.

During these phases, the architect will issue invoices for a portion of the design fee, usually based on the projected total cost of construction. Typically, there will be a retainer fee of from $1,000 to $10,000, depending on the size of the project. This retainer will be a credit toward the total fee due at project completion. The programming fee will be about 5 percent of the total fee. Schematic fees are about 15 percent. Preliminary fees are usually 20 percent. Working drawings will be about 40 percent of the total fee. Bidding will be another 5 percent. Construction administration will use up the remaining 15 percent.

Staying in Touch and Organized

Communication is easy today. Nearly everyone in business carries a cell phone. Most have constant access to text messages. Your problem may be ensuring that team members make good use of the communication tools available.

Every CM contractor has authority to set communication policy for the job — the routing of questions, requests, comments, notices, approvals, reports, memos, submittals, decisions, invoices and payments. Requests from contractors, vendors and government authorities should be routed through you. If a decision by the owner is required, you'll refer the request. Responses from the owner should be routed through you to ensure that you stay informed on what's happening.

It's a good idea to keep a current call list and the text message list for each project. Distribute that list on a need-to-know basis. We'll have more to say about this in Chapter 7.

As the CM contractor, you'll want to be informed of nearly all issues that affect quality, schedule or cost. You'll decide which issues have to be referred to the owner for decision. But make it clear to contractors and vendors that only the owner has authority to order materials or contract for services.

Retaining Records

Set up a filing system for each project. Maintain another file of correspondence, accounting records and reports. Good records win cases. If you get into a dispute with the owner, a contractor or vendor, the most persuasive proof will always be a written record made at the time in question. If you sense the possibility of a misunderstanding based on an oral communication, get it in writing. A text message (email) is quick and easy and makes good proof when preserved in a digital archive.

All contracts should be kept in another file. This file should include signed contracts, changes to those contracts, and everything incorporated by reference into those contracts. That can include specifications, estimates, bids, detail drawings, general and special conditions, shop drawings, letters of acceptance, the project schedule, and exhibits.

It's especially useful to have these documents available on or near the construction site. You'll be surprised at how often you'll need to refer to these files. If there's a secure job shack or locker on-site, that's usually a good place to store plans and contract documents. If not, use a metal file box that fits in the tool carrier in the bed of your truck. The file of correspondence, accounting records and reports are best kept at your office off-site.

Your Contract for Pre-Construction Services

The option *Manage design services* in *Construction Contract Writer* adds to your agreement the services discussed in this chapter:

> *Contractor will develop an understanding of the Project as conceived by Owner and will recommend design and engineering professionals best able to meet the criteria for the Project set by Owner. Contractor will Work with design personnel approved by Owner to create a plan for the Project which meets the needs of Owner. If required, Contractor will recommend alternative building materials, systems and construction details which allow completion of the Project within the time and budget established by Owner. The plan developed under direction of the Contractor will include working drawings and specifications adequate to construct the Project in compliance with applicable law, regulations and ordinances.*

You won't see the option *Manage design services* unless you select the fifth option under *Plans for this Project*. Selecting any other option is confirmation that plans have already been prepared and that the CM contractor won't participate in development of the plans.

One other caution to note. At present, six states (CA, IL, MA, NV, PA and TN) require that contracts for home improvement work (including pools) show the total cost in dollars and cents. In those six states, *Construction*

Managing Pre-Construction Services

You won't see the option *Manage design services* unless you select the fifth option under *Plans for this Project.*

Selecting any other option is confirmation that plans have already been prepared and that the CM contractor won't participate in development of the plans.

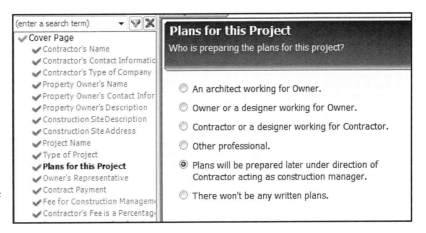

Construction Contract Writer
CM Contract Options

Click to include management of design services in the scope of work.

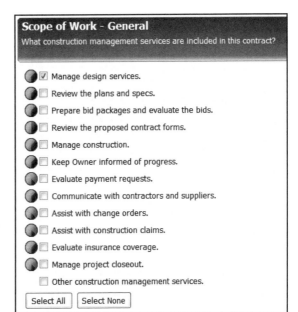

Contract Writer won't draft one contract that covers both home improvement design services and construction. To contract for home improvement design services in one of those six states, write two contracts, the first for managing design services at a fixed price. When design is complete, write a second CM contract at a fixed price (or a guaranteed maximum price) for the remainder of the work.

Where We're Headed

Not every CM contract will include design services. *ECI* (early contractor involvement) isn't the norm for CM contracts. More typically, CM contractors don't get involved with a construction project until design is nearing completion. The next chapter picks up work at a point where conventional general contractors will feel right at home — with a review of the plans.

Chapter 4

Reviewing the Plans and Specs

IN THE LAST CHAPTER we identified the four principal design phases:

- *Programming.* The scope of work is defined in words.
- *Schematic design.* The written scope of work is turned into drawings.
- *The preliminary design.* Hard line drawings show exact areas.
- *Working drawings.* The plans that will be used by construction crews.

If your task includes coordinating any of these four phases, you have an opportunity to set and enforce design limits — including budget, schedule and size. Set limits on height, configuration, shape, construction methods and green design. If you're involved during any design phase, begin reviewing the plans weekly. Don't delay plan review until the working drawings are complete. That's way too late to get serious about design decisions. If possible, get a contractor involved on Day 1, pricing every pencil line as it's drawn. Spend at least an hour each day poring through the design work. The best time find an error is the first time an error shows up on paper. Expect your reward (and the owner's) to be worth several times the effort.

> **Why Plan Review is Important**
>
> *The opportunity for mistakes on the plans is greater today than at any time in my career.* – William D. Mitchell
>
> In days gone by, plans were drawn by hand. A designer who wanted a particular detail had to physically draw that detail line by line, considering each line and word as it was drawn. That isn't what happens today. With computer-aided drafting, a designer who wants a specific detail simply finds that detail in some old plan set. He clicks to copy and clicks again to paste that detail into the current drawing. Very little thinking is required. A designer doesn't even have to understand what's getting pasted. The result: All sorts of nonsense can end up in your plans, like store-front details for a stud wall in the plans for a concrete block warehouse. Miss a drafting-import error like that and you're headed for a train wreck during construction.

If you step into a project after working drawings are complete, the options are more limited. Plan checking at this point is like a search and rescue mission — trying to limit the potential losses that design exuberance can bring into the project. Still, no matter when you assume responsibility, there will be good opportunities to influence design, as we'll describe in this chapter.

Plan Checking

Most serious questions about budget and schedule should have been resolved early in the design process. As you get closer to working drawings, the focus of plan review shifts to anticipating problems. We've heard contractors claim that the cost of fixing a design error on the plans is about 1 percent of the cost of fixing the same problem once construction has started. We can endorse that sentiment.

Never rely on the building department to find errors and omissions in your plans. The plans examiner will find some mistakes in plan check. But plan check for code compliance is entirely different from the plan review recommended in this chapter. Believe it or not, there's nothing in the building code that guarantees something on the plans can actually be built or, if built, will serve the intended purpose.

You've probably heard stories about construction companies that bid work at cost and then make a bundle on changes. To some, that may be good business. But it's clearly *not* good policy. As a rule of thumb, change orders required to fix errors or omissions in the plans shouldn't add more than 3 percent to the construction cost. A good set of plans won't require more than one change order for each five pages of plans and specs.

But even a single change order can be too much. I saw a new school being built with covered exterior walkways supported by round poured concrete columns. The job included more than a hundred columns. As drawn on the plans and as built by the contractor, the tops of these columns were about 6 inches too short to meet the roof elevation. Many meetings and weeks later, the columns were finally the right height. The owner absorbed a six-figure loss. The contractor made a bundle — and had probably recognized the problem the first time he saw the plans!

Plan Review

Don't put the job out to bid until you're satisfied that the plans are as complete and accurate as humanly possible. Even the best, most professional A/E firms make mistakes. At a recent symposium, one of the participants described what he found when reviewing plans for a large hospital.

> *We had a firm of national repute doing the design of a project, a hospital, with construction cost estimated to exceed $100,000,000. They had to resubmit their 100 percent design drawings and then their working drawings four times before they were right. Even in the last submission, there were still beams missing girders and unsized structural members.*

There's a reason architecture is referred to as a *practice*. If you want something done right the first time, buy a toaster. The first few hundred toasters to roll off the assembly line probably got tossed out. Only when assembly is perfect are toasters sold to the public. In the design and construction business, you get the first unit built, whether perfect or otherwise. If the plans omit 20 percent of what's required for the job, staying within the owner's budget will be a struggle all day, every day.

Never just skim through a set of plans. There are all kinds of notes, and you should read every last one. The next few pages will help you spot errors when reviewing plans and specs. We've included examples with many points commonly overlooked when planning construction projects. Every one of these items has some impact on cost. Leave any of these unresolved until work starts and you've got a charge for extra work and maybe a bitter dispute.

Is this a complete list? Certainly not. Thousands of additional items could be added. But we feel these hit the high points. Everything covered was either an expensive mistake on a construction project, or about to become an expensive mistake. Contractors tend to remember expensive mistakes. You will too. Use these points as a guide when reviewing any set of plans.

> **The Importance of Plan Review**
>
> Here are the best reasons I know for careful plan checking:
>
> - If a contractor makes a mistake, such as installing the wrong type of door, he installs the right door at his own expense.
>
> - If an architect makes a mistake, such as specifying the wrong type of door, who pays? The owner pays — both to have the old door removed and the right door installed. That's why they call it an architectural *practice*.
>
> Another example: Every electrical plan includes a floor plan with circuit wiring. Circuits are usually run in conduit, either EMT, flex, PVC, GRS or IMC. On residential projects, sheathed conduit (Romex) can usually be substituted for conduit. Romex cable costs about half of what EMT conduit and wire will cost. The specs or a schedule on the electrical plan should identify the type of conductor (CU or AL) and the type of conduit, if any. If your plans omit this little detail, will the building department catch the error? Probably — but maybe not. If the problem isn't found in plan review, who's going to find the mistake? That's easy. The building inspector will find the error during rough electrical inspection. At that point, you've got an expensive problem. The project has to be rewired. And the next question will be: Who pays?
>
> - The A/E left something off the plans. But the A/E's contract probably disclaims liability for omissions like this.
>
> - The building department didn't catch the error. But the city isn't going to write a check to cover the loss.
>
> - The electrical sub was just following the plans. He bid the job as economically as possible and demands to be paid for rewiring the job.
>
> - Certainly it wasn't the owner's fault. The owner didn't do anything wrong.
>
> But, of course, the owner is going to pay. If you're involved in a loss like this, you're going to remember the project for many years.

The Six Cs of Plan Review

Plan review can be broken into six distinct categories. Each of these categories can be examined separately. In fact, that's exactly what we recommend. The first review should look for anything that's missing. Are the plans complete? Then do a second review to be sure plans and specs are consistent. Continue through the list, one "C" at a time.

Complete — Are the plans an accurate and thorough representation of the proposed project? Have principal design features been approved by the owner and all concerned? Is every detail that's called out in a bubble actually drawn somewhere? Are enough sections shown to define the project?

Consistent — Is each sheet of the plans consistent with every other plan sheet? Are the plans consistent with the specs and with standards cited in the

"If I told you what kind of day I had, it would take a week."

plans and specs? Find yourself a light table and overlay sheets on each other, one sheet at a time. Does the ceiling plan match the floor plan? Do the light fixtures fall where air conditioning grilles are already located? Do plumbing vents on the roof pass through the heating and cooling units?

Clear — Ambiguity in plans is the enemy of productivity. Good plans require the least improvisation. Don't leave room for several interpretations of what's required. When there's doubt, you're going to get the cheapest and least desirable option. Clear, unambiguous plans don't require issuing multiple addenda in response to questions from prospective bidders. Avoid unnecessarily complex plan details. A contractor who has trouble understanding some detail will factor more contingency into the bid price. Plan details are tedious work for a designer, and they're usually the last part of a plan set to be prepared. The more details an architect has to figure out, the more time it will take. So they tend to limit the amount of detailing. But a set of plans light on details will be heavy on change orders. Every project requires a balance between the cost of design and the expense of change orders.

Correct — It takes time to fix errors and distribute corrections. Serious errors may require a second trip through plan review at your local building department. To avoid delay, get it right the first time. Use weekly meetings to identify errors. When you get a progress set of plans from the architect, do a weekly plan review while the contractor is doing the weekly cost estimate. When you find a mistake, have the architect make corrections before the next weekly meeting. It's a whole lot easier to check six sheets each week than 60 sheets after 10 weeks of drafting. Regular plan review helps cut the job down to size.

Constructible — Designers and architects are perfectly capable of drawing plans that can't be followed at any reasonable cost. Be sure the plans allow enough space for trade contractors to complete their work. For example, does the plenum above a ceiling have space for both duct and conduit? In a residence, be sure framing details leave enough room for drain lines. Even if space is tight, your plumber, electrician or HVAC contractor will find a way. But you may not like the result. When in doubt, ask for a three-dimensional plan. View the plans from the installer's perspective. I call this the *hand test*. Try to visualize getting your hands inside the space shown on the detail. Could you install the screws or finish and paint the drywall in that space? If not, don't expect an installer to do what you wouldn't even attempt.

Whenever possible, favor materials that are readily available and that can be installed using conventional construction techniques. Avoid work that can be performed by only a few highly-compensated specialists. Contractors are (quite reasonably) reluctant to bid projects they consider exotic. For example, almost all construction materials come in square shapes. Almost nothing comes round. Building anything round requires forcing square materials into round shapes. It can be done. But there's usually a lot of waste. That runs up costs with little benefit. If your architect's first design includes circles, triangles, trapezoids or a rhombus, think about switching architects.

Cost — A/E firms aren't always sensitive to cost issues — either construction cost or the cost of maintenance. For example, the same symposium participant quoted earlier found good ways to cut costs during plan review on the same hospital project.

> *The design called for more than two dozen different shades of interior paint. Can you imagine trying to maintain that kind of an inventory or match these shades over a period of years?*
>
> *We have learned that a simple metal railing along the hallway not only offer patients a handhold, but saves thousands of dollars in maintenance costs by preventing gurneys and other equipment from bumping into the wall. The designers had not thought of that.*

When not contrary to the owner's intent, consider replacing brand name specs with material grade specs. A brand name in the specs is a command — required no matter what the cost or delay. Specifying grades rather than brands permits substitutions. In the case of hardware, the words *Baldwin 85360* in the specifications permits installation of only a Baldwin 85360 lockset. Using the words *Baldwin 85360 or equal* permits substitution of a similar ("equal") product. Where substitutions are allowed, the specs should identify how equivalents will be approved, probably during the bid process.

> *"Manufacturers of architectural-grade windows make it easy for architects to specify their products."*

In some cases, specifying one manufacturer "or equal" may not be enough. Baldwin is somebody and "or equal" is nobody. To make the "or equal" clause work, it may be better to say "Baldwin, Quickset, or equal." Now the two named suppliers are competitors and "or equal" opens the door to more choices. *Equal* to a single model from a single supplier is a very precise standard. Some could argue that *equal* means nearly identical. Naming two suppliers "or equal" is a much easier standard to satisfy.

Another example: Some window manufacturers promote their products to architectural offices. Manufacturers of architectural-grade windows make it easy for architects to specify their products. In nearly every case, specifying windows by brand and model number will guarantee that the owner gets a premium window at a premium price. Other window manufacturers may offer similar styles and grades at considerably lower prices. If the owner is price-conscious or is eager to prevent construction delay due to long lead times, consider specification by grade rather than brand. This is known as a *performance spec* — how the window is supposed to perform. For example, a performance spec for a window might identify how the window resists wind and rain, inhibits UV rays and controls temperature.

Review for Biddability

When you're done with the big picture (the six Cs), it's time to begin checking details. Do the plans and specs answer every question a contractor will have about what's required? Can what's shown on the plans actually be built? Begin by checking design issues.

Dimensions (height, width, depth) — To build any project successfully the dimensions have to be spot on, otherwise you'll end up in the hot seat in the owner's office. To ensure that all the designer's dimensions are accurate you'll have to crawl through them, one by one. Begin by checking the length of one outside wall. Add up the chain of dimensions along the entire wall. Then check the full length of the opposite outside wall. Checking to the center of walls doesn't work well because that makes the framer do the math in his head, quickly — like what's half of a 3⅝ stud? Be sure the two dimension chains match. If they don't, it's called a *bust* in the dimensions. When the contractor lays out the foundation, the forms won't close. If the contractor forces the forms to close, nothing from the foundation to the roof will fit together right. Everything will be out of square.

Drafting programs usually don't volunteer the sum of dimensions across an entire wall. So it's easy to miss a bust in the dimensions. Suppose the sum at the top of the plan is 74 feet 9 inches. Then later, the sum of the dimension across the bottom, or the opposite wall, comes out 74 feet 11 inches. A draftsman might not notice the 2-inch difference. Why the extra 2 inches on the bottom wall? Probably because the number of rooms at the top of the plan and the bottom of the plan are different. I call that *computer dimensional plaque buildup*. Repeat a ⅛-inch difference a dozen times or more, and the total will be off by several inches — and so will your budget.

The draftsman should have instructed the computer to make the chain of dimensions 74 feet 9 inches across both the top and bottom walls. That doesn't happen by default. The draftsman has to give the instruction. Computer-aided drafting is amazing. But no drafting program is any better than the draftsman. Mistakes like this happen all the time. Crawl through the plans with a fine-tooth comb and you'll find plenty of inconsistencies.

When you've checked the length of all opposite outside walls, begin checking dimensions of inside walls. When you're sure the dimensions of the inside walls are right, start checking wall heights.

Materials (composition, grade, color, texture, manufacturer) — Every material identified on the plans warrants a description. Most materials will be described in the specs or in a detail drawing. For example, where rebars are specified, do those rebars require epoxy coating? Nuts and bolts should be identified by tensile strength and include torque settings. Sod and seeding should be specified by seed mix.

Inconsistencies (between and among the plans, specs and other contract documents) — Specifications are usually created by copying from one document and pasting into another — often without a careful reading. Plan details may be pasted from a plan archive without thought about how the detail integrates with the rest of the project. Be sure that what's pasted into the plans and specs makes sense in the context of those plans and specs. For example, on five of the sheets, the designer's callout shows *brick*. On the sixth, the callout shows *concrete block*. On the remaining sheets, the callout specifies *concrete* or *masonry*. Verify that callouts on all sheets are consistent. *Siding* can refer to any type of siding. Lap siding isn't cove siding, or composite siding, or aluminum siding, or wood siding — but they're all siding.

Plan notes — These are prone to errors. Check for conflicts with plan details and specifications. Plan notes can refer to the specifications but shouldn't duplicate information in the specifications. Rough-sawn beams and sidings aren't the same as re-sawn beams and sidings. The difference can be tens of thousands of dollars. If a note says "See engineer," is there anything relevant on the engineer's drawings to refer to? The engineer may have no idea what the designer is referring to and neither will the contractor. Plan coordination simply broke down or never got done.

Schedules — Pay special attention to schedules. For example, door, window and hardware schedules should answer every question about the materials to order. I worked on casino project with schedules that included over 10,000 doors and windows. Door and window schedules went on for dozens of sheets and were revised almost daily. Don't take any schedule lightly. Even a small error on a schedule can delay a project for weeks. It's a serious problem if 200 doors show up on a jobsite hinged on the wrong side. The plans should leave no doubt about which door, window or piece of hardware gets installed at each location. If paint and surface preparation are covered in a schedule, pay special attention to requirements for metal surfaces. The specs should leave no room for interpretation.

Electrical work — Is the type of conduit (GRS, EMT, flex) identified? If all wire is intended to be copper, do the plans or specs say that? Are motor starters required for the HVAC system? Check any point where dissimilar metals join. For example, any time black iron pipe joins copper, you've built a battery. Current will flow between the two metals, gradually converting both metals into something that resembles corrosion and not a building. A galvanic fitting should be specified. Does fuse alignment in switchgear have to be at the centerline of the enclosure? Is the primary disconnect switch top feed only? Do terminal lugs match the conductor in size and amperage? Do the specs require that the fire alarm installer be state certified? If wire will be installed in existing duct, are pull wires present in that duct? Will temporary lighting be required? If so, the specs should require proper supports. Do the plans show proper grounding? Are disconnect switches lockable? Do you want to specify a procedure for pulling wire in conduit?

Large electrical rooms require code-compliant exiting. Two exits may be required. Most electrical switchgear requires a clear space in front of the panel. Allowing for that clear space can morph a 10- by 10-foot electrical room into a room that's 15 by 20 feet or larger. Cooling electrical rooms can be another big issue, especially computer rooms. Many computer rooms require huge bundles of cable run in a raised plenum floor. Access to those cable bundles can bring another set of headaches.

Mechanical work — Is environmentally-friendly Freon required? (R12, R22 and R123 are no longer permitted in the U.S.) Is the fire-sprinkler test station piped to an exterior drain? Does the NFPA sprinkler system require certification? Do the plans show water pressure assumptions for the design of fire sprinklers? Do the specs permit attachment of mechanical system components to the roof deck (rather than structural components)? Do motorized valves show a wiring diagram? Do the plans show pipe sizes for cleanouts? If any pipe wrap is used, will a field test be required? Does the piping system require cathodic protection? Are cast iron valves and fittings permitted? Are extended valve stems required so valve handles can operate clear of insulation? What pressure test is required for piping systems? Are traps sized properly for fan coil units and air handlers? Do specs for piping identify what pipe has to be welded and what pipe can be threaded? Do the plans show fire sprinklers installed above high-voltage equipment? Are test instruments required to have a recent calibration certificate? Will smoke alarms and fire sprinklers have to be tied to a central alarm panel and the local fire department? That's generally required on public buildings. Will heat detectors be required in attics, concealed spaces and stairwells?

Non-Design Issues

The most common oversights on plans will usually be non-design issues. Check each of the following:

The site — Are you on the correct site? I've had an owner tell me to build on a lot that was next door to the lot he actually owned. Are property lines and the net buildable area depicted on the plans? Are survey markers in place and identified on the plans? Do site conditions conform to what's represented on the plan? Does the plan show an elevation reference you can verify, such as a benchmark or height above the curb? Are utility lines and access points identified? Will a high water table affect construction? Will temporary fencing or silt barriers be required? If backfill and compaction are required, do the specs include compaction values? Will you need to cut an access road to the site and haul in base rock to stabilize the surface in wet weather? Will the site need retaining walls not shown on any plans?

Site clearing — Is the area to be cleared well-defined? Will any trees or shrubs have to be relocated and maintained during construction? Are underground obstructions clearly identified?

Removal — If building components or landscaping have to be removed, will the contractor have to deal with either hazardous materials or structural pests? Do the specs identify horizontal and vertical limits for removal? What's the plan for collection of debris on-site and disposal of debris off-site? Are there limits or restrictions on disposal of debris? Who has salvage rights to materials? Will it be necessary to protect adjacent surfaces during demolition? Will some proportion of waste materials have to be recycled (such as for LEED certification)? Will the project require relocation of utility lines? Toxins on old factory and gas station sites are real deal killers. Did you ever wonder why old corner gas station sites are so slow to be redeveloped? A fuel tank leaking into the ground for 30 years can require a clean up that costs well over a million dollars.

Limitations — Will there be practical or legal limits on construction activity? For example, is there an enforced noise ordinance? Are there limits on work hours? Is overhead space available for lifting equipment, if required? Will weather conditions make work impractical for long periods? Will a permit be required for use of laser equipment? Space will always be required for storage of equipment and materials. Is that space defined on the plans? Do the plans show access routes for the public, equipment and delivery of materials?

Traffic plan — Are there restrictions on parking? If a public right-of-way will be obstructed during construction, will a permit be required? What temporary signage or warning devices will be required?

Drainage — Is there an erosion control plan? How will construction affect the existing drainage? How will new drainage connect with existing drains? Nearly all communities regulate drainage into public storm drains. Most urban areas have two drainage systems, one for storm water runoff and another for sewage. In some urban areas, the same piping handles both storm water and sewage. That's a major problem in heavy rain. Sewage is likely to spill into rivers or the ocean. In communities without natural drainage, huge storm water retention basins may be required, taking up much of the available site.

> *"Many jobs require special inspections or tests that won't be performed by the building inspector."*

Inspections — Many jobs require special inspections or tests that won't be performed by the building inspector. Examples include concrete mix design, metal tensile strength, and soils and aggregate base consolidation. Who schedules and pays for these tests? Be sure test standards are clear and that a testing organization is available locally to perform each test. Many building departments now require that special tests be listed on a testing and inspection schedule on the front sheet of the plans.

Safety — Is welding required? If so, verify proper handling of weld fumes. If space on-site is limited, will a storage yard be available? What personnel protection gear will be required (respirators)? What fall-protection will be

"We told them it's because we believe in hiring locals first, but actually, it was the only way we could stay within the owner's budget."

required? Any use of enamel paint or lacquer on-site will require special precautions.

Warranty — What warranties are required? If the owner expects warranty protection by a manufacturer, is the contractor obligated to deliver warranty documents? If the contractor will provide a warranty, how will the owner invoke terms of that warranty? Is that plan feasible?

Finally — When you find an error or omission, look for similar mistakes of the same type elsewhere in the plans. Errors tend to run in packs. For example, schedules are error-prone because writing schedules is slow, boring work. When you find a bust in dimensioning, look for more busts on other plan sheets, including schedule sheets.

Accumulate a list of questions until your review is complete. Keep that list by plan sheet or trade. When done, forward your list to the A/E for resolution. If you've identified valid errors or omissions, the A/E firm will be more than willing to make the correction. Resist the temptation to make a change on the plans or specs yourself. When plans are stamped with an A/E firm's seal and license number, that firm is taking responsibility for everything on the plans. The A/E firm has the right to expect that everything on those plans reflects their work. An A/E firm that permits changes to the plans without their direct supervision will be subject to discipline by the state board.

Cost Estimates

Keeping costs within the owner's budget is a huge part of every CM contractor's task. Until you have firm bids (offers from qualified trade contractors), the job will probably require several cost estimates. If you've agreed to provide successive cost estimates, you'll have to develop one or more of the estimates that follow.

We've recommended doing cost estimates weekly during the design process. Even if you've come on board after the design was complete, develop a plan to avoid surprises. Don't rely on an architect's estimates, as they don't buy labor and materials on a daily basis and don't know current costs. Contractors do, so only estimates done by qualified contractors should be used.

The contractor's estimate should be equivalent to a bid for the job — what the contractor would charge for doing the work. If a contractor wouldn't sign a contract to build it for that price, don't trust the estimate. Of course, every estimate is only an opinion of cost at a specific point in time. Each of the estimates described in the next section includes a typical range of accuracy. It's fair to apply an expected range of accuracy when making any estimate before working plans are available.

Types of Estimates

At project inception — This estimate is sometimes called a *feasibility study*. It's intended to answer the question, "Is this project a good idea?" This estimate is usually based on square foot costs and can be prepared without knowing much about site conditions, the plans, materials to be used, or the contractors selected. All you need is the approximate square footage. Many square foot cost reference manuals are available. Most cover new construction. Remodeling and repair work will be much more difficult to estimate on a square foot basis. But it's safe to assume that remodeling and repair work will cost more per square foot than new construction. Whether new construction, repair or remodeling, this is the first estimate the owner expects to see and will help establish the owner's budget. A project inception estimate (1) sets parameters for preliminary design work, and (2) becomes the basis for discussion of financing options with the proposed lender. If the project is considered feasible, planning will continue. Expect actual costs to be as much as 20 percent less to 100 percent more than this type of estimate.

Start a project estimate with a walk on the site. I invite the owner, contractor, architect, engineer and staff to join me. Everybody gets a yellow notepad and is encouraged to make copious notes — which I copy and distribute to everyone involved. Then we throw back in a big conference room and have a skull session on the scope of work, budget and schedule. Play project scope, budget and schedule off against each other and everyone in the room. By the end of the day, you'll know what's in, what's out, and if the project makes economic sense.

On completion of preliminary design work — This estimate will probably be based on square footage. But you can now consider quality of construction, building shape and design features, choice of materials, site development costs and site conditions. Include specific allowances for finish materials, installed equipment and yard improvements. When selecting materials, remember that heavy materials, like concrete and masonry, usually cost more than lighter materials. Supporting concrete or masonry at the second or third floor level requires large steel beams which add more weight. That requires larger footings and more rebar. Light and durable products will be easier on an owner's budget.

When approved by the owner, this estimate becomes a baseline for all design work that follows. Changes in design that require a change in this estimate need approval from the owner.

I develop this kind of an estimate weekly during the design phase. Don't wait until the designer has nearly finished work. By that time, design work is hopelessly downstream. Backtracking will be painful, expensive, and schedule busting. Expect actual costs to be as much as 15 percent less to 75 percent more than this type of estimate.

On completion of all design work — This estimate requires a labor and material takeoff for principal project components (concrete, lumber, steel, windows, doors, roof), but will probably be prepared by reference to published cost data adjusted to the site with percentages. This estimate should identify likely costs due to unusual site conditions, weather (based on the start date), fees imposed by government agencies (permits, charges, remediation), charges for utility hookup, and LEED certification, if required. Expect actual costs to be as much as 10 percent less to 50 percent more than this type of estimate. At this point, the owner should have enough information available to get a commitment from the proposed lender.

When contract documents have been drafted — At this point, the project is ready to bid. This will be the last project estimate prepared by a CM contractor. Refinements at this stage will probably include recent changes in labor and material costs and may include quotes from likely material suppliers on principal components. Expect actual costs to be as much as 10 percent less to 25 percent more than this type of estimate.

There won't be any surprises if a contractor prepares an estimate every seven days during the design phase. Every successive estimate needs only seven days worth of updating. You may not even need to put the job out to bid. The contractor has been pricing multiple subs for weeks and already has the lowest prices.

You have an obligation to share these estimates with the owner. If costs may exceed budget, and if the project could be put in jeopardy by cost overruns, intervene on behalf of the owner. Suggest to the design team alternate materials or designs that could help keep costs within the owner's budget.

We're not going any further on the subject of cost estimates. This isn't a book about estimating construction costs. Many good references are available, including several by Craftsman. See the last few pages of this manual.

Pulling the Permit

The permit issuing process is different in every city and county. Every building department has the right to establish their own rules, and they all have. But some generalizations apply in nearly all jurisdictions. What follows is an introduction to what you can expect.

The prime contractor is usually responsible for pulling the permit. But there's no reason why an appropriately-licensed CM contractor (or even an A/E firm) can't apply for a permit. Licensed mechanical and electrical contractors usually prepare drawings for mechanical and electrical work and will usually apply for permits on their work. Simple plans that don't require plan review may be approved right at the front counter in your building department office. Larger projects may spend weeks (or even months) in plan review. The larger the project, the more problems you're going to have with permits. You can find consultants on the Web who specialize in applying for permits and getting permits issued.

Plan review is just one part of the permit-issuing process. Your project also has to comply with state laws, regulations and local ordinances administered by other government agencies. For example, if zoning issues are involved, your project will need approval of the planning department. If fire sprinklers are required, the fire marshal's office will review the plans for compliance with the fire code. If the project is a restaurant, expect that the health department will want to check for compliance with the health code. In any case, the building department won't issue a permit until your project is in full compliance with all government regulations. A checklist on the building department's website should help you identify approvals needed by other government agencies. If in doubt, call your building department for more guidance.

Where to Start

Your application for a building permit on a small project can start at the building department's website. Search for the city or county name and "building department." That site will probably include a blank application for a permit and a list of required documents. Don't walk into the building department office without all the required documents and a completed application form.

For a larger project, you'll usually start at the planning department. They check for land use, zoning, setbacks, height compliance and conditional use permits. If your project is controlled by a homeowners association, you'll have to get HOA approval before filing with the planning department. When you have approval of the HOA and the planning department, file with the building department. While your plans are in the building department, you'll likely be required to submit plans to the public works department for a review of sewer, water, storm drainage, streets, curbs, sidewalks, bus turnouts, road widening and street lighting. The utility companies serving your parcel may have to certify that water pressure, power, sewer, gas, cable, internet access and telephone service will be available at the site. The fire marshal's office will review fire hydrants, fire water pressure and flow and your landscaping plans for vegetation that could fuel a fire. The department of education may charge school impact fees. That cost can be anywhere from $1 to $5 a square foot.

Once you've made your way through all of that misery, haul your various approvals back to the building department and pull your building permit.

If your plans are on paper, the technician at the front counter of the building department will do a quick scan for required basic information — names and contact numbers and enough detail to permit evaluation of code compliance — and then accept the plans for review.

On larger projects in urban locations, you'll usually have to set up an appointment with an intake technician at the building department. At that

Building inspector's night at home.

meeting, the technician will do a preliminary review. If your plans have all the required calculations, including Title 24 energy efficiency information if required, the plans will be accepted for plan check. If any information is missing, plans won't be accepted. You'll have to come back later when your plans are complete. Building departments started refusing plans during the building boom of the 1990s. Before that time, designers tried to shorten the time required for approval by filing for a permit when plans were about half done — knowing full well that the plans would sit in a back room bin for four to six weeks. Before the plans reached the top of the plan check pile, the designer would substitute plans that had a chance of getting approved.

If your plans are on a CD or DVD, the technician will probably check the disks for compatibility with their computers. If your project is very simple, like reroofing, the permit technician may take your check for the fee and issue a permit right on the spot. Otherwise, the technician will calculate the fee, which is usually based on an assumed value per square foot. Plan check fees are set by ordinance and will usually be equal to about 65 percent of the building permit fee. The building permit fee will usually be about 2 percent of the cost of construction up to $100,000, and 1 percent of the cost of construction over $1,000,000. The technician will then accept your check, issue an application (tracking) number and pass your documents to the plans examiners. A large building department will have specialists for structural concrete, electrical work, plumbing, HVAC, etc. Many smaller building department offices have plans examiners who work on a fee basis — usually at a different location.

Most plans examiners rely on a checklist when reviewing plans. The building department probably publishes on their website the checklists used by plans examiners in that office. Your A/E firm almost certainly has a copy of the appropriate checklist and will probably run down that checklist before considering the plans 100 percent complete. If your building department doesn't offer a copy of their plan review checklist, Lynn Underwood's book *Building Code Compliance for Contractors & Inspectors* has an excellent 8-page checklist for residential construction based on the *International Residential Code*. An inexpensive download of that book is available at http://Costbook.com. The city of Novi, Michigan offers a good plan review checklist for commercial construction. Search on the Web for "City of Novi Commercial Plan Review Checklist." The most comprehensive plan review checklist is available for about $50 from the International Code Conference at http://ICCsafe.com.

Making Corrections

Depending on project size, location, time of year and the state of the economy, a week to as much as 12 weeks may pass before you or your A/E firm hears from the plans examiner. If a correction is required, the examiner will issue a comment letter listing problems that have to be resolved. The content of that letter will be filed under the assigned tracking number to simplify processing the resubmission. The plans (on paper or disk) will be returned for correction — usually citing specific code sections. If only minor changes are required, there won't be an additional fee on resubmission.

You or your A/E firm are welcome to call the plans examiner to discuss any issue in the comment letter. Most plans examiners will be ready to suggest the best way to get your plans approved. We don't advise splitting hairs with the examiner. Most plans examiners are seasoned experts at code interpretation and compliance. It's seldom worth anyone's time to prove that the examiner is wrong. Nor is it a valid argument that a detail rejected on your plans this week was approved last week by some other plans examiner at another office.

Show respect for the plans examiner's position. But if you feel the examiner's interpretation of the code is wrong, you're welcome to initiate an appeal — assuming you have several weeks to wait while the appeal is processed.

When the permit is issued, you'll get a list of required inspections, usually 12 or more for a larger project. We'll discuss these inspections in Chapter 7, *Managing the Day-to-Day Work*. All cities and counties require that the permit be posted in plain view on-site. On residential projects, nail the permit to the wall inside the garage. On a large project, hang the permit on the wall in the job trailer.

While plans are in review, use the time to develop a project schedule. Once the permit is issued, it's time to begin following that schedule.

The First Schedule

Have you ever had the electrician show up to begin rough wiring before ceiling joists were installed? Have you ever had to keep the plumber waiting while framers relocate a partition wall? Whose fault is it when the inspector shows up for final inspection before the electricians are finished?

Every construction project deserves a schedule. Each task completed is a link in the chain that follows some prior link and precedes a later link. Scheduling is just making sure that each link falls neatly in sequence so work can go from start to finish by the most direct and most profitable route possible.

A schedule is simply a list of tasks to be executed in order — from top to bottom. Scheduling with pencil and paper works fine — until something changes. Then you need a good eraser. Some CM contractors schedule with a chalkboard or a grease pencil. That's OK too. But you'll need a big board to handle a large project. Years ago, most construction schedules were made with bar charts and arrow diagrams. Today, most scheduling is done with a

"The cement truck is due at noon and I forgot to schedule the formwork."

computer. Expect to spend $500 to $1,000 for a flexible, comprehensive, professional-grade scheduling program. If you don't want to get that deep into scheduling, pencil and paper or a stack of 3 x 5 cards may serve nearly as well.

Any scheduling tool you select has to be flexible. Schedules change. Revising the schedule should be a simple process. If the project lasts more than a month, you'll want to revise the schedule at least monthly. Identify both work completed on time and work not completed as planned. Consider ways to get back on schedule without increasing costs and without sacrificing quality. Then revise the schedule based on work yet to be done and what you've learned about the performance of the trade contractors assigned.

Our advice is to take scheduling very seriously. The most successful CM contractors know more about their plans, budget and schedule than anyone on the job. You don't review plans, budget and schedule monthly, weekly or even daily. You live with them hourly. The plans, budget and schedule should be on your night stand when you go to bed and on your breakfast table when you get up in the morning.

Schedules and estimates are closely related. A detailed labor and material cost estimate will identify how many manhours and how much material will be required. When you have a manhour estimate, divide by the crew size to calculate how long a task will take.

Advantages of a Good Schedule

Scheduling makes it easier to manage a construction project. A good schedule makes prompt completion more likely and reduces idle or wasted time. A good schedule should also reduce or eliminate resource emergencies — discovery that essential labor or materials aren't available when needed.

Coordination and scheduling are important responsibilities for every CM contractor. On the smallest projects, one person can keep track of nearly everything that has to happen. On a larger project, a calendar or checklist may be all that's needed to remind you of important dates. As work becomes more complex and timing becomes more critical, someone has to begin laying out work schedules and charting progress.

This isn't a book on scheduling. We're not going to describe in detail how to use a scheduling program. Plenty of other good references are available. But regardless of the scheduling tool you select, certain concepts (terms) will be useful.

Schedule Components

You've probably heard the terms *CPM* (critical path method) and *PERT* (Program Evaluation and Review Technique). There's no need to understand the difference between these or other approaches to scheduling. They're the same in many respects and can be referred to collectively as *echeloning tools*. Each identifies when a task can begin, how long it should take, and when it should end. Detailed schedules show both a sequence of trades on the job and the sequence of tasks to be performed by each trade. With a detailed schedule, you can set both delivery times for materials and arrival times for trade contractors.

No matter the type of schedule, you'll need some basic information to begin:

- the item description
- the unit of measure (such as CY or MBF)
- the quantity of work to be completed
- where this task fits into the job sequence
- the expected starting date
- time required to complete each item of work
- the forecast completion date
- float (delay) time allowed between tasks

Every project schedule begins with a list of the work elements for each trade. Next, assemble those elements into a logical sequence. Excavation has to come before foundations, wall construction has to precede roof framing, sub-base and base preparation come before paving. Identify the first day of work as Day 1. Every day thereafter is assigned a number in sequence.

Of course, you don't have to wait until one trade is done before beginning the next trade. For example, plumbing and wiring rough-in can start before the last roofing tile is laid. But any time work transitions from one trade to another, allow time for correction and repairs. For example, framers may have to add bracing when plumbers are finished cutting drain lines through joists.

Be especially sensitive to durations that don't involve construction trades. For example, it takes time to get permits. Inspections aren't always completed on the day scheduled. And, of course, nearly every job gets delayed at least a few days by weather. Delay is nearly inevitable. That's why every realistic schedule includes *float*, the time a particular task can be delayed without delaying the entire project.

Until the project is put out to bid, focus more on project duration than on scheduling trade contractors. The owner needs an answer to a simple question, "How soon can I move in?" That, of course, depends on when construction begins. Early in the planning process, the occupancy date may be no more than an educated guess. Once work starts, you'll use the job schedule to prepare progress reports. Each successive report should reduce uncertainty about the occupancy date. Regular progress reports are part of a CM contractor's responsibility, as we'll explain in Chapter 8.

That's as far as we intend to go on the subject of scheduling. Any scheduling program you select will include specific instructions. Most scheduling programs, including Microsoft *Project*, are available on a free trial basis. Don't buy until you're satisfied that the program is a good value and will meet your needs.

The Next Step

Once plan review is complete, it's time to begin collecting bids for the work. That's where we're headed next: Preparing Bid Packages.

Chapter 5

Preparing Bid Packages & Evaluating the Bids

WHEN YOU'VE REVIEWED THE plans and specs as described in the previous chapter, it's time to select a contractor or contractors to do the work. The first question will be: Who gets to bid?

On public works projects, state and federal law usually require that bid competitions be open to everyone qualified. There's no such requirement on private jobs. You can award the job on any arbitrary grounds you select. We don't recommend that, but we do recommend that you consult with the owner before making any decision about the award. The owner may have strong feelings on this issue and will want to participate in the decision. These are the most common choices:

> ➢ Negotiate the contract with a single contractor.
> ➢ Ask for bids from a selected list of contractors.
> ➢ Open the bidding to any contractor qualified to do the work.

There's nothing wrong with negotiating trade contracts. The CM contractor mentioned in Chapter 1 does just fine using the same trade contractors on job after job. In fact, he very seldom puts any work out to bid. He's comfortable using a small number of specialists, trusts them to do nothing but first class work, and is perfectly content with the prices they charge. His is a business based on trust and mutual respect. Not every CM contractor is that fortunate.

If you've been working with a contractor during the design phase, and if that contractor's estimates have met the budget requirements of the owner, awarding the job to that contractor on a negotiated basis makes perfect sense. In fact, it would be questionable practice at this point to put the job out to bid. Get one more estimate from the contractor you've been working with and sign an agreement.

If you're completely comfortable with awarding work at a negotiated price, you know exactly what has to happen next. You don't need bid packages or forms. You can skip this chapter and go on to Chapter 6 or 7.

If you plan to open bidding to more than a select few contractors, plenty can go wrong during the bidding process. Using the forms in this chapter and following our recommendations could save more than a little grief.

Before deciding who gets to bid, remember that every CM contractor has an obligation to work in the best interest of the owner. Open competitive bidding eliminates the most obvious opportunities for favoritism and self-dealing. That's why competitive bidding is often required on public works projects.

The fact that you're asking for bids on the project doesn't necessarily mean you've ruled out awarding the contract on a time-and-materials (*cost plus*) basis. Even T&M contracts can be awarded to the lowest responsible bidder. If your choice is T&M trade contracts, you'll ask for bids based on hourly labor rates, the contractor's fee and, perhaps, a guaranteed maximum price.

Public vs. Private Contracts

Not all projects done by government agencies require competitive bidding. Many government agencies have authority to award smaller contracts on a negotiated basis. Even larger projects may not require competitive bids.

Some public agencies in California have managed an end-run around competitive bidding statutes with what's usually called a *lease-lease* contract. The owner, a school district for example, may want a new facility on an existing site without going through the traditional design, bid, build process. The owner leases the site and existing buildings to the contractor for a specific price. The contractor is now the owner's tenant. Next, the contractor subleases the site and buildings back to the owner at a specific price. Once the leases are in place, the contractor can rebuild the site and buildings as a private, not a public project. And, as it happens, what the contractor wants to build is exactly what the school district needs. The owner pays the contractor under a negotiated contract, as though this were a private project. Once construction is complete, the leases are cancelled. The owner occupies the new facility and the contractor moves on to other work.

So far, no court has invalidated a lease-lease contract. It may happen eventually, as it's clearly an end-run. But in the meantime, I've managed more than a $100,000,000 worth of lease-lease construction in California. This form of contracting actually started in the public sector and has, in the last few years, moved into educational facilities. They pay state-mandated prevailing wages to placate the unions.

The principal defect in the public bidding process is that you lose control over who gets the contract. Too often, you end up with a contractor who can post a bond, but:

> ➤ Made a big mistake when bidding the job and now hopes to make up for the loss by leaning across your desk, drinking your coffee, and arguing for change orders.
> ➤ Has never prepared a schedule of values and has to be led by the hand through the process.
> ➤ Can create a pretty good schedule, but is hopeless when it comes to staying on that schedule.
> ➤ Is never on-site. Instead, his brother-in-law runs the job because he's a stockbroker and just got laid off.

As a CM contractor, I prefer to do business either with a contractor I know personally or one I've pre-qualified thoroughly. When I can do that, we're going to get something built very quickly, and the work will be done right. My favorite job is a design/build, lease-lease back negotiated contract with a contractor hired at the same time as the architects. That type of project nearly runs itself. I'll be done in half the time it takes to build a traditional design, bid, redesign, build, and then litigate project.

The Bid Package

For the remainder of this chapter, we're going to assume that you and the owner have decided to open bidding to more than a select few contractors. As part of the invitation to bid, you'll distribute plans and specs to those interested in bidding. Every prospective bidder will also want to see the contract they'll be asked to sign. That's entirely reasonable. What the contract says will influence both the cost of completing the job and the risk in doing the work. We recommend that you make draft contracts available to prospective bidders at the same time you make plans and specifications available. We'll suggest how to prepare these trade contracts in the next chapter.

Recruiting Trade Contractors

Finding contractors willing to bid on your project won't be particularly difficult. Finding contractors *qualified* to bid on your project may be harder. The obvious choice will be trade contractors you already know to be qualified. To expand the search, ask your A/E team to recommend prospective bidders known to be both active in the area and qualified to undertake the work. You may decide to limit the search to a select list of highly-recommended contractors. Open bidding will nearly always attract more prospects, including many with only marginal qualifications. Regardless of your choice, understand the accepted wisdom in the construction industry: The more bidders, the lower the contract price; the lower the contract price, the greater the risk of default. You have an obligation to deliver good value for the owner — that means finding the best balance between cost and risk.

If you decide on competitive bidding open to all, it's good practice to screen prospects early in the process. You're not going to recommend a contractor who's obviously unqualified to take on the work — regardless of the price. We recommend using the Pre-Qualification Statement in Figure 5-1 to identify prospective bidders who wouldn't be selected to do the work even if they're the lowest bidder.

"Actually, McCrackle, I'm glad to know you have a G.E.D. But we were hoping for a contractor with O.J.T."

Getting the Word Out

Try to pre-qualify at least six bidders. Expect about half to show up at bid time. If you start with only three prospective bidders, you could end up with only one bid. If you have only one bidder on bid day, call off the bid, do more solicitation legwork, then rebid the project 30 days later. Don't count as bidders any subcontractors who show up on the job walk. The only serious bidders are general contractors.

To improve exposure to qualified bidders, consider making your plans available at a plan room in your community. Most projects offered in plan rooms are larger jobs, often public works construction. But plans for smaller jobs will still be welcome in most plan rooms. The A/E firm or designer who prepared your plans may have a suggestion on the plan room best for your job.

> *"If you start with only three prospective bidders, you could end up with only one bid."*

If your plans are in digital form, consider using a digital (electronic) plan room. Plans for many projects are distributed either on the Web or on disk. Any prospective bidder can view the plans on their own computer. If printed plans are necessary, a local FedEx Kinko's may be able to print the plans from disk or by download off the Web.

Pre-Qualification Statement

(Page 1 of 4)

Please complete this form and return the form with a current financial statement to the contract manager listed in Box 2. If you need extra space to answer any question, attach a continuation sheet headed with a box number from where the answer is continued.

Statement Submitted By (Box 1) This statement is intended to qualify the submitting party to bid on a contract for construction of the project identified in Box 4 and is submitted by:

Name of the person completing this form

Company name

Address

Address

Voice phone

Fax number

Email address

License or registration number

State of incorporation (if incorporated)

The Contract Manager (Box 2) When completed, return this form to:

Name

Address

Address

Voice phone

Fax number

Email address

Date (Box 3) Date this statement is completed.

Date

The Project (Box 4)

General description

Location

Short job name

Experience (Box 5) List five projects completed successfully in the last five years by the company identified in Box 1. Favor projects which are similar to the project which appears in Box 4.

Project 1

General description (use 2 lines if needed)

Location

Date completed

Reference (name of a contact for this project)

Reference (phone number or email address)

Project 2

General description (use 2 lines if needed)

Location

Date completed

Reference (name of a contact for this project)

Reference (phone number or email address)

Figure 5-1
Pre-Qualification Statement

Project 3

_____

General description (use 2 lines if needed)

Location

Date completed

Reference (name of a contact for this project)

Reference (phone number or email address)

Project 4

_____

General description (use 2 lines if needed)

Location

Date completed

Reference (name of a contact for this project)

Reference (phone number or email address)

Project 5

_____

General description (use 2 lines if needed)

Location

Date completed

Reference (name of a contact for this project)

Reference (phone number or email address)

Work Currently in Progress (Box 6) List projects currently in progress by the company identified in Box 1.

Project 1

_____

General description (use 2 lines if needed)

Location

Estimated completion date/Percent currently complete

Reference (name of a contact for this project)

Reference (phone number or email address)

Project 2

_____

General description (use 2 lines if needed)

Location

Estimated completion date/Percent currently complete

Reference (name of a contact for this project)

Reference (phone number or email address)

Figure 5-1, continued
Pre-Qualification Statement

Project 3

_____

General description (use 2 lines if needed)

_____

Location

_____

Estimated completion date/Percent currently complete

_____

Reference (name of a contact for this project)

_____

Reference (phone number or email address)

Project 4

_____

General description (use 2 lines if needed)

_____

Location

_____

Estimated completion date/Percent currently complete

_____

Reference (name of a contact for this project)

_____

Reference (phone number or email address)

Project 5

_____

General description (use 2 lines if needed)

_____

Location

_____

Estimated completion date/Percent currently complete

Project 5, continued

_____

Reference (name of a contact for this project)

_____

Reference (phone number or email address)

Value of Work in Progress (Box 7) Estimate the total value of all contracts currently in progress and the bonding capacity of the company identified in Box 1.

$ _____
Value of all contracts

$ _____
Total bonding capacity

Disputes (Box 8) List two of the current or most recent lawsuits, mediations, arbitrations or disputes in which the company in Box 1 has been involved.

Dispute 1

_____

General description (use 2 lines if needed)

_____

Month and year when the dispute originated

_____

Amount in controversy

_____

If the dispute has been resolved, the month and year of resolution

Dispute 2

_____

General description (use 2 lines if needed)

_____

Month and year when the dispute originated

_____

Amount in controversy

_____

If the dispute has been resolved, the month and year of resolution

Figure 5-1, continued
Pre-Qualification Statement

(Page 4 of 4)

Trade References (Box 9) List a contact and phone number or email address for five of the company's largest suppliers.

Supplier #1 name

Contact name

Phone number or email address

Supplier #2 name

Contact name

Phone number or email address

Supplier #3 name

Contact name

Phone number or email address

Supplier #4 name

Contact name

Phone number or email address

Supplier #5 name

Contact name

Phone number or email address

Bank Reference (Box 10) List a contact and phone number or email address for the company's primary banking relationship.

Bonding Company Reference (Box 11) List a contact and phone number or email address for the company's primary bond underwriter.

Name of bank

Contact name

Phone number or email address

Name of underwriter

Contact name

Phone number or email address

Signature (Box 12) My signature affirms that this statement is true and correct to the best of my knowledge.

Signature

Printed name

Company name

Date

Title

Please return this form and a current financial statement for this company to the address in Box 2.

Figure 5-1, continued
Pre-Qualification Statement

Many architects (and some contractors) prefer traditional paper plans. The scale on electronic plans may not match precisely the scale on paper plans. And electronic plans present a hazard not associated with paper plans: Digital files can be changed and copied. Most architects don't allow their work be copied freely or changed at will by access to a digital file. If digital files are available for bidding, they'll more than likely be read-only files that can't be printed or revised.

Avoid distributing partial plan sets. It's important that every prospective bidder have access to *all* contract documents. Everyone bidding should be responsible for coordinating their work with other trades. That helps ensure bids that are comparable. A trade contractor who received only a partial plan set is going to bid only those partial plans. A mistake by a contractor caused by incomplete plans will become your mistake and require correction at the owner's expense. If some trade contractors need less than full plan sets, let the primary trade contractors accept responsibility for dividing up the work. If you feel it will be helpful, offer to make a full set of contract documents available to trade contractors at your office or some other convenient location.

If you plan to open the bidding to all qualified contractors, consider listing the project with the local daily construction publication. Most communities have an online or printed listing of projects open for bid. Again, most projects listed will be public works or larger commercial projects. But nearly any project open for competitive bid will be welcome. To find your local construction reporter, search on the Web for *daily construction reporter* or *construction project leads*. Add the name of your state or county to limit the search to your community.

The Invitation to Bid

Don't begin soliciting bids until you have the necessary documents ready. It makes a poor impression on prospective bidders when you're not prepared to deliver a full set of contract documents when the project is announced. Expect bids to be more competitive when qualified trade contractors recognize that they're dealing with an experienced professional ready to do business.

Every bid competition has the potential to turn into an acrimonious dispute. Selecting a bid is also a rejection of other bids. Every losing bidder has invested time and effort in expectation that you'll deal honestly and fairly with all bidders and award the contract on grounds announced in advance. Do anything else and you're inviting legitimate complaints or even a lawsuit. If you're not prepared to award the contract to the lowest qualified bidder, it may be better to negotiate a deal with the contractor you prefer.

Bidding for public works contracts is governed by state or federal law. The agency or contracting officer may have very little discretion in making the award. That's not true for the award of contracts for privately-financed construction. There are almost no legal restrictions on private bid competitions. But the best way to retain the respect and confidence of prospective bidders is to conduct a fair, even-handed, open and well-organized competition.

Every open invitation to bid for the award of a significant private construction contract should include at least the following:

- Invitation to Bid
- the Bid Form
- Rules for Bidding

The Invitation to Bid will refer to *contract documents* which describe exactly what's going to be built. These contract documents will include the contract, plans and specifications, and may include other items, such as detail drawings or samples.

Addenda

A perfect Invitation to Bid, when read together with other contract documents, will answer every potential question and resolve every likely dispute. You may never see an Invitation to Bid as perfect as that. More often, you'll have to issue *addenda* to the plans or the specs or the Bid Forms, or even to the proposed contract.

Most addenda will be issued in response to questions received from prospective bidders. Unfortunately, the need for addenda doesn't usually become obvious until the last few days (or even the last few hours) before bids are due. That can make it nearly impossible to distribute the addenda to all bidders. Obviously, it's important that all bids be based on the same information and the same assumptions. That's not going to be the case if some bidders receive all addenda and others don't.

Be sure addenda are numbered, dated and distributed to all prospective bidders. Box 4 of the Bid Form, which is the second part of the Invitation to Bid, requires each bidder to list the number of addenda received. Any bid which omits consideration of a material addendum is subject to disqualification.

Resist the temptation to respond to a legitimate question from one prospective bidder without providing the same information to all bidders. If something is wrong or misleading or omitted from your Invitation to Bid, you have an obligation to issue addenda to all prospective bidders. If time doesn't permit distribution of addenda to all prospective bidders before the bid deadline, the bid opening may have to be postponed while addenda are distributed. Your Invitation to Bid should cover that possibility.

Consider setting a drop-dead date for questions that could require an addendum. A good deadline would be two or three days before the bid opening. If you get an important question after that date and expect an addendum will be required, move the bid date back several days or weeks and notify all bidders in writing.

The Invitation to Bid should identify whether or not the plans and specs have been approved by the building department (see Box 8). If plans aren't yet approved, set up a plan check allowance in the contract to cover the cost of any changes required by the building department. Also, identify whether you've filed an application for service with all the public utilities involved. On larger projects, getting connected to the utility grid can be very expensive, and it can take months for them to get their engineering work done — a process you don't control.

Think of Everything

In the last chapter we covered common errors and omissions in building plans. Those items were intended to help you anticipate problems before they became expensive mistakes. The Invitation to Bid, shown in Figure 5-2, at the end of this chapter, will serve the same purpose. Of course, no form will resolve every conceivable issue in a bid competition, but we believe our blank Invitation to Bid form is very complete. In fact, it may be overkill for your project. If so, just extract from this form only what's needed on your jobs. But at least review each of the topics covered. Many, if not most, of the topics covered here will be relevant to every significant construction project.

The Pre-bid Conference

A serious prospective bidder will want to visit the construction site before preparing a bid. This visit is usually called a *job walk* and is good estimating practice. An alert construction estimator who walks a proposed site will identify many conditions that can influence cost. Assuming the site is vacant ground and isn't protected by a fence or gate, prospective bidders are free to do their job walk any time they want.

But there are good reasons to combine a job walk with a pre-bid conference conducted by the construction manager. It simply saves time to have all prospective bidders present when you answer questions about the job. You won't have to notify bidders individually of each material issue raised during the conference. Be sure you have a sign-in sheet available at the conference and ask everyone present to sign in, listing their company, phone number, email address, whether they are a contractor or subcontractor, and the date.

Schedule the conference for about two weeks after documents are first available and no later than two weeks before the bid due date. To help estimate the number of bidders on bid day, make the pre-bid conference mandatory. If you suspect the bid day turnout will be light, prompt other bidders to participate. There's nothing worse than having only one bidder show up on bid day. It makes you look amateurish to the owner.

It's good practice to have someone take minutes of the conference, or produce an audio or video recording. Offer to distribute the minutes or recording to anyone who couldn't attend personally. Keep another copy in your file as defense against claims made later about undisclosed site conditions.

Encourage those who plan to attend the pre-bid conference to review the plans and specs before the conference begins. Those with the most information about the project will ask the best questions. Also, encourage everyone attending to bring a set of the contract documents. Prospective bidders will refer to plans and specs many times during the conference.

Disadvantages

The obvious disadvantage of a scheduled job walk and pre-bid conference is that it gives bidders a good idea about the level of competition. Most of those at the conference will recognize representatives from other construction companies. That could be an advantage if bidders understand that competition will be intense. If the pre-bid conference isn't well-attended, some will leave with the opposite impression. In any case, any meeting of prospective bidders also creates an opportunity for collusive bidding, never a welcome prospect for any owner. Despite the risk, most construction managers prefer a scheduled pre-bid conference at or near the jobsite.

Conducting the Pre-Bid Conference

The goal of every pre-bid conference is to build confidence and reduce the opportunity for misunderstanding among prospective bidders. Accurate information helps eliminate the perceived risk. If the project includes unusual features or potential construction problems, be candid and proactive. Address the most sensitive issues before questions are asked. Bidders need to understand that you're not trying to hide anything. For example, describe how any toxic materials will be handled. Is the owner going to take care of their removal or is the bidder going to inherit that problem? If the latter is the choice, then how is the cost of testing and removal being accounted for in the bid? Are there limits on the area where toxic materials have to be removed?

> *"A good pre-bid conference will develop at least a few issues you hadn't considered."*

In the last chapter, you did a careful review of the plans. Prospective bidders shouldn't have many questions about the plans and specs. But prospective bidders will have questions the first time they see the construction site. The list of questions starting on this page will help you identify points about the site that may need to be addressed at the pre-bid conference.

A good pre-bid conference will develop at least a few issues you hadn't considered. When attendees are well-prepared, you're going to get a great free plan check — and may have to issue addenda or even re-submit plans to the building department.

Questions about the Site

Prospective bidders may have questions such as the following about the construction site:

General information
- Will any part of the site be occupied during construction?
- Will other (separate) contractors have access to the site?
- What signs, barricades, traffic control or temporary lighting will be required?
- Are any noise or dust control limits enforced?
- What work is implied though not shown on the plans and specs?

Site
- What are the limits of the construction area?
- Is there a survey showing benchmarks, utilities and structures?
- What area will be available for material storage (lay-down area)?
- Will any recorded (or unrecorded) easement affect construction?
- Will there be limits on access to the site?
- Are locations of utility lines well-defined?
- What changes will be needed in routing existing utilities?
- What's known about requirements for utility hookup?
- Will an easement be required to connect with utility lines?
- What provision has been made for temporary power, water, sanitary, cable, TV and phones?
- Which existing fences, drains, landscaping or improvements have to be preserved?

Soils
- What's known or suspected about soil conditions?
- Is there a soils test? Is it less than 12 months old?

Excavation
- Can excess excavation material be spread on-site?
- Is structural fill work involved?

Demolition
- Is there a demolition plan that identifies the extent of demolition?
- What demolition will be required before construction begins?
- Are any hazardous materials known to be on-site?
- Are there any limits on removal of hazardous materials?

Safety
- What safety measures are needed to protect work crews?
- What safety training or licenses will be required for work crews?
- What safety measures are needed to protect the public?

Security
- What security measures will be enforced?
- Do any security measures place limits on access by the trades?
- What security measures are needed at night and on weekends?

If the contract documents include a proposed schedule or projected completion date, you're likely to have scheduling questions. Be ready to discuss:

- each of the proposed milestones
- any liquidated damage provision for missing a milestone
- any bonus for early completion
- assumptions about the weather, regulatory approval and work by others
- the lead time for materials and equipment to be ordered by the owner
- beneficial occupancy (occupancy before substantial completion)

Awarding the Contract

On public works projects, bidders are usually invited to attend the bid opening. The contracting officer will announce the name of each bidder and the bid price. No negotiation is permitted at a bid opening. It's customary to announce the *apparent* low bidder when all bids have been opened. But the contract shouldn't be awarded until the three lowest bids have been checked carefully for compliance with the bidding rules.

Box 6 of the Bid Form at the end of the chapter allows 10 days for award of the contract after bids are opened. This gives the construction manager an opportunity to confirm that the apparent winner has complied with the Invitation to Bid and the bidding rules and is qualified. It also allows time for the following considerations.

> If there's an error in the low bid (such as failure to acknowledge all addenda), give the winning bidder an opportunity to withdraw the bid.

> If the low bid doesn't conform to the invitation or the rules, then notify the apparent winner that the bid has been disqualified.

> If the low bid conforms to the invitation and rules but includes qualifications, give the bidder an opportunity to remove these qualifications.

> If the apparent winner doesn't qualify as responsible, meet with that bidder to resolve any issues. If unsuccessful, advise the bidder of reasons for disqualification.

> If the low bid is more than 10 percent below the next lowest bid, give the low bidder an opportunity to confirm the accuracy of the bid and withdraw if an error is found.

> Ask the apparent winning bidder about workload. If the value of the current project is more than 50 percent of the value of work completed in the previous year, the current project may be too large for this construction company. If work in progress has increased by 50 percent or more over the last year, the contractor may be taking on too much work. If you identify workload problems, give the apparent winner an opportunity to withdraw the bid.

When you're satisfied that the three lowest bidders have complied and are qualified, you can use an Excel (or similar) spreadsheet to compare bid details. Or order a copy of *Construction Forms for Contractors*, which includes a ready-made Bid Comparison Sheet on the CD included. Find it on the order form bound into the back of this book. When listed in three adjacent columns, you should have no trouble selecting the most attractive offer. Share this spreadsheet with the property owner. Summarize for the owner what you've learned from any dialog you've had with the apparent winning bidder. Then make a recommendation on the award of the contract.

When you agree on the award, have the owner sign both copies of the Bid Form (Box 11). Return one signed Bid Form to the winning bidder. The owner should also sign duplicate originals of the construction contract. Send both originals to the winning bidder for signing. One of the duplicate contracts should be returned to the owner for filing. Box 6 of the Bid Form gives the winning bidder five days to sign these duplicate original contracts. During this period, the contractor selected should submit certificates of insurance, performance bond forms and other documents required by the contract.

Moving On

Throughout this chapter we've referred to the *Contract Documents*. These include the plans, specifications, and, of course, the construction contract. Until now, we haven't looked carefully at the contract. That's the subject we'll cover in Chapter 6.

Invitation to Bid

(Page 1 of 8)

Qualified contractors are invited to respond to this Invitation to Bid by returning the Bid Form. The contract for this work will be awarded on a competitive basis as described in this invitation and the Rules for Bidding.

The Contract Manager (Box 1) This Invitation to Bid is issued by

_____ _____
Name   Address

_____ _____
Address   Voice phone

_____ _____
Fax number   Email address

Please use any of these contact numbers if you have a question about this project or any Contract Document. Notify the Contract Manager promptly if you intend to bid. That ensures your name appears on the list to receive addenda which modify the Contract Documents.

Date (Box 2) Date this Invitation to Bid was issued.

_____
Date

The Project (Box 3)

_____
General description

_____
Location

_____
Short name used to identify this project

_____
Project owner/representative

Delivery of Bids (Box 4) To respond to this invitation, complete the Bid Form (or an accurate facsimile) and deliver by the Bid Due Date *(check all that apply)*.

- ☐ To the address in Box 1 in a sealed envelope
- ☐ By fax to the number in Box 1
- ☐ By email to the email address in Box 1
- ☐ To the following address in a sealed envelope

_____ _____
Address   Address

Bid Due Date (Box 5) The Bid Form (or an accurate facsimile) must be received by

_____ _____
Date   Time

Any response to this Invitation to Bid which is incomplete or is received after the Bid Due Date or which does not conform to the Rules for Bidding will not be considered. Modification or withdrawal of a bid after the Bid Due Date will be subject to the Rules for Bidding. Bids will be evaluated without discussions.

Contract Price (Box 6) This Invitation to Bid requires a response which includes *(check all that apply)*.

- ☐ A contract price (exclusive of any alternate bid, unit price or contract allowance).
- ☐ One or more alternate bids as identified in the Contract Documents.
- ☐ One or more unit prices as identified in the Contract Documents.
- ☐ One or more contract allowances, owner's allowance and stipulated contingencies, as identified in the Contract Documents.

Figure 5-2
Invitation to Bid

(Page 2 of 8)

Bonds and Insurance (Box 7) Each Bid Form must be accompanied by *(check each that applies)*:

☐ A bid bond as required by the Contract Documents

☐ Certificates of insurance as required by the Contract Documents

☐ Other _____

_____

Contract Documents (Box 8) To respond to this Invitation to Bid, you will need the Contract Documents, which consist of this Invitation to Bid, the Bid Form, the Rules for Bidding, addenda issued after this invitation was distributed and *(check each that applies)*

☐ **Plans**

_____
Dated

Consisting of _____ sheet(s)

_____
Prepared by

_____
Date of last change

☐ Plans were approved on _____

☐ Plans have not yet been approved

☐ **Specifications**

_____
Dated

Consisting of _____ sheet(s)

_____
Prepared by

_____
Date of last change

☐ Specs were approved on _____

☐ Specs have not yet been approved

☐ **Detail drawings**

_____
Dated

Consisting of _____ sheet(s)

_____
Prepared by

_____
Date of last change

☐ Detail drawings were approved on _____

☐ Detail drawings have not yet been approved

☐ **Construction contract**

_____
Dated

Consisting of _____ sheet(s)

_____
Prepared by

_____
Further described as

Figure 5-2, continued
Invitation to Bid

Contract Documents (Box 8), Continued

☐ **Site information** (soils report, hazmat, etc.)

_____ _____
Title(s)   Dated

Consisting of _____ sheet(s) _____
                              Further described as

_____
Prepared by

☐ **Project manual**

_____ Consisting of _____ sheet(s)
Dated

_____ _____
Prepared by   Further described as

☐ **Shop drawings**

_____ Consisting of _____ sheet(s)
Dated

_____ _____
Prepared by   Further described as

☐ **Samples**

_____ _____
Prepared by   Further described as

☐ **Schedule or time limitations**

_____ Consisting of _____ sheet(s)
Dated

_____ _____
Prepared by   Further described as

☐ **Applications for service with utility companies** (check the one that applies)

  ☐ Have been filed   ☐ Have *not* been filed yet

☐ **Other**

_____ _____
Prepared by   Further described as

Permits and utilities permitted on _____

Figure 5-2, continued
Invitation to Bid

(Page 4 of 8)

Distribution of Contract Documents (Box 9)

The Contract Documents are available at _____

Charge for the Contract Documents (Box 10) The charge for these Contract Documents is:

$ _____

☐ This charge is a deposit which will be refunded within 10 days after return of the full set of Contract Documents in usable condition.

Pre-Bid Conference (Box 11) A conference for all prospective bidders will explain site access, scheduling, testing and other job requirements.

_____   _____
Date and time   Location

☐ This pre-bid conference is required for all prospective bidders.

☐ This pre-bid conference is recommended but **not required**. If you cannot attend this conference in person, ask the Contract Manager to arrange monitoring of this conference by phone, audio or video recording.

Figure 5-2, continued
Invitation to Bid

Bid Form

(Page 5 of 8)

To have your bid considered, complete, sign and deliver <u>two copies</u> of this form to the address on the Invitation to Bid. Bids which do not comply with the Invitation to Bid and the Rules for Bidding will not be considered.

Bid Submitted By (Box 1) This bid is an offer to enter into a contract for construction of the project identified in Box 3 and is submitted by:

_____ _____
Individual name   Company name

_____ _____
Address   Address

_____ _____
Voice phone   Fax number

_____ _____
Email address   License or registration number

_____
State of incorporation (if incorporated)

Date of this Bid (Box 2) This offer is made on

_____
Date

The Project (Box 3) This bid is a response to the Invitation to Bid

_____
Invitation to Bid date

_____
Issued by

_____
Project name

_____
Owner

Acknowledgement of Addenda (Box 4) The undersigned acknowledges receipt of the following addenda (changes) to the Invitation to Bid or Contract Documents

_____ _____
Number of addenda received   Date of last addendum

_____
Description of last addendum

Bonding and Insurance (Box 5) This Bid Form is accompanied by *(check each that applies)*

☐ The bid bond required by the Contract Documents

☐ The certificates of insurance required by the Contract Documents

Bid Price (Box 6) If this offer is accepted within 10 days after the Bid Due Date (or such postponements as may be allowed in the Rules for Bidding), the undersigned will, within 5 days thereafter, sign the contract referenced in the Invitation to Bid.

$ _____

Bid price (exclusive of any alternate bid, unit price or contract allowance)

Figure 5-2, continued
Invitation to Bid

Alternate Price (Box 7) Alternate price or prices (if required by the Contract Documents)

$ _____ _____

$ _____ _____

$ _____ _____

Unit Price (Box 8) Unit price or prices (if required by the Contract Documents)

$ _____ _____

$ _____ _____

$ _____ _____

Contract Allowance (Box 9) Contract allowance or allowances (if required by the Contract Documents)

$ _____ _____

$ _____ _____

$ _____ _____

Signature of Bidder (Box 10) This offer is submitted by

_____   _____
Signature   Printed name

_____   _____
Title   Company name

Acceptance by Owner (Box 11) The signature below indicates acceptance of this offer by the owner and creates a binding contract without further action by either party.

_____   _____
Signature   Printed name

_____   _____
Title   Company name

_____
Date accepted

Figure 5-2, continued
Invitation to Bid

Rules for Bidding

(Page 7 of 8)

These rules have been adopted to simplify award of the contract and to ensure that all bidders have equal opportunity. The rules checked in the paragraphs that follow apply to this Invitation to Bid and the Bid Form. Bids which do not comply with these rules will not be considered.

_____

Date of the Invitation to Bid

_____

Project name

_____

Project owner

☐ **True Name**
Bidder must provide in Box 1 of the Bid Form a true name and address, including street, city, state, and zip. If incorporated, the bidder must show the state of incorporation. The bidder (or authorized agent) must sign Box 10 of the Bid Form.

☐ **Notice of Intent to Bid**
Bidder must give notice to the Contract Manager upon receipt of the Contract Documents. Notification will ensure that you receive addenda to the Contract Documents issued after this Invitation to Bid. Failure to give this notice may make your bid non-conforming.

☐ **Withdrawal of Bid**
Any bid may be withdrawn until the Bid Due Date but not thereafter. A bidder who has withdrawn a bid may submit another bid so long as that bid complies in all respects with the Invitation to Bid.

☐ **Acceptance of Offer**
The Contract Manager intends to recommend that a contract be awarded to the lowest responsible bidder whose offer conforms to the Invitation to Bid and to these Rules for Bidding. However, the Contract Manager may recommend rejection of all bids or acceptance of other than the lowest bid. Only the project owner has authority to accept or reject any offer received from a bidder. If requested by the Contract Manager, bidder agrees to provide a statement of qualifications showing that the bidder is qualified and prepared to complete the work required by the Contract Documents.

☐ **Alternate Bids, Unit Prices and Contract Allowances**
If the Contract Documents require alternate bids, unit prices or contract allowances, the Bid Form must show the required alternate bids, unit prices or contract allowances. Failure to submit a required bid price will disqualify the bid. If the Contract Documents do not require an alternate bid, unit price or contract allowance, any alternate bid, unit price or contract allowance on the Bid Form will be disregarded.

☐ **Pre-bid Conference**
A ☐ mandatory ☐ optional pre-bid conference has been scheduled to brief prospective bidders on project requirements. Check the Invitation to Bid for the time and place of this conference.

☐ **Addenda**
The Contract Manager reserves the option to make changes in the Invitation to Bid or Contract Documents or the Bid Due Date until bids are opened. Addenda will be sent to every prospective bidder who has given notice of intent to bid. Addenda will be numbered sequentially and dated. If there's a break in the sequence number of addenda, assume you do not have a full set of the Contract Documents. Contact the Contract Manager immediately.

☐ **Questions**
The Contract Manager will be available to answer questions. But responses by the Contract Manager will be limited to topics which help explain or clarify the intention of the Invitation to Bid, Contract Documents or the Rules for Bidding. Don't expect the Contract Manager to respond to any question which might give any prospective bidder a competitive advantage.

☐ **The Award**
When the award has been made and the contract signed, details about the winning bid will be released at the request of anyone who submitted a bid on the project. If no award is made, bids will not be released.

Figure 5-2, continued
Invitation to Bid

☐ **Insurance and Bonds**
If a bid guarantee is required by the Contract Documents, bidders are required to furnish the guarantee prescribed. Failure to provide the required guarantee with the Bid Form will be grounds to reject the bid.

☐ **Minor Irregularities**
A minor irregularity or clerical error is an immaterial matter of form rather than substance and does not disqualify a bid from consideration. A defect or variation is immaterial when the effect on price, quantity, quality or delivery is a negligible part of the agreement. Examples include failure of a bidder to (1) Return the number of copies of signed bids required by the invitation; (2) Sign the bid if it is clear that the bidder intends to be bound by the unsigned bid; (3) Acknowledge receipt of an addenda to a Contract Document if the addenda has a negligible effect on price, quantity, quality, or delivery.

☐ **Illegible Bids**
Any bid that includes an essential term which is unreadable, incomplete or omitted will be rejected. Transmission of bids by electronic means is at the risk of the bidder. Erasures and strikeouts will be considered so long as the change is clearly legible and initialed by the party signing the Bid Form. If space on the Bid Form does not permit a complete response, attach a continuation sheet headed with the box number from where the response is continued.

☐ **Other rules for bidding**

☐ **Mistakes in Bids**
The Contract Manager will correct obvious clerical mistakes after verification by the bidder. For example, misplacement of a decimal point or use of a decimal point rather than a comma would be an obvious clerical mistake.

☐ **Postponements**
If it becomes necessary to reschedule the Bid Due Date, the Contract Manager will issue an addendum to the Invitation to Bid.

☐ **Substitutions**
Make requests for substitution of materials to the Contract Manager at least 7 days before the Bid Due Date.

☐ **Bidding on a Portion of the Work**
If the Contract Documents permit a bid on separate parts of the project, a bidder may show a price for any separate part. Or a bidder may indicate that an award of less than all parts of the project will not be accepted.

☐ **Sealed Envelopes**
Except when submitted by electronic means, bids must be enclosed in a sealed envelope.

Figure 5-2, continued
Invitation to Bid

Chapter 6

Reviewing the Trade Contracts

MOST CONSTRUCTION CONTRACTORS HAVE a favorite contract they use over and over again. If you have a favorite contract, use this chapter to evaluate that contract. You'll probably discover a few additional points you should cover in your contract. If you don't have a favorite set of contracts, use this chapter to help you develop contracts that meet your needs.

Nearly all contractors understand the advantage in writing their own agreements. The party who controls the contract also controls the bottom line. That's accepted wisdom in the construction industry. Contracts offer protection. The best way to get that protection is with a contract you draft.

You've got a major advantage in contract negotiations if you're good at drafting contracts:

- You can make quick and easy contract revisions.
- Your printed contract looks highly professional.
- You can be confident the contract complies with state law.
- You can deliver a contract almost instantly by email.
- You can deliver the contract in a format that can't be easily changed.

Drafting construction contracts requires an understanding of both law and construction. We suggest that it's easier for you to master a few basic principles of contract law than it is to teach an attorney the essentials of construction.

Most attorneys have very little understanding of the design and construction process. What comes first, the rebar or the concrete? What's the difference between rough and finish electrical? What comes naturally to an architect or a builder is alien territory — almost a foreign language — to most lawyers.

Use your knowledge of the design and construction process to create tailor-made contracts. No magic words are needed when writing contracts. Plain language works best. Leave no doubt about what's intended. Define what the owner has to do and what the contractor has to do — the project, the schedule, and the money. Be sure there's a way into a project and a way out of the project if things don't go as planned.

As a CM contractor, you need two types of contracts.

1. **Construction management contracts.** These are pure consulting contracts. You don't plan to install any materials or write any subcontracts. All purchasing and contracting will be done in the name of the property owner. Still, these CM contracts have to comply with state law for the type of work done and value of the work. Visit http://PaperContracting.com to download free CM contracts legal in each of the 50 states and the District of Columbia.

2. **Trade contracts**. This chapter covers the contracts you distribute with bid packages. The property owner signs these contracts once the contractors are selected. These contracts also have to comply with state law in all respects. Visit http://Construction-Contract.net to download free construction trade contracts legal in each of the 50 states and the District of Columbia.

In the previous chapter, *Preparing Bid Packages and Evaluating the Bids*, we assumed that you had a trade contract prepared and ready to distribute to prospective bidders. This chapter will provide the background you need to draft good trade contracts.

Essentials in Every Construction Contract

What's in the contract depends on the job and the state where work will be done. But every construction contract has to cover several key points. The first of these usually falls on what we'll call the *cover page*: Names and numbers for the parties (the construction contractor and the property owner), identification of the construction site, and the contract price. Lines for signatures will be on the last page of the agreement. You'll usually sign duplicate originals. Both are considered to be *the contract* for legal purposes.

Between the cover page and the signature page, nearly every construction contract will cover at least five broad subjects.

"It's a good contract legally, but your jargon's weak."

Details about the work:
- Identify the plans and specs
- Who owns the plans
- Project location
- Contract amount
- Start and finish dates

Scope of the work:
- What's included and excluded
- Contingencies
- Materials and equipment
- Supplied by the owner
- Obligations of the contractor
- Obligations of the property owner

Payment:
- Is this a cost-plus (time and material) contract?
- Will there be an initial payment?
- The dates and amount due for progress payments
- What happens if payment is delayed?
- Grounds for withholding all or part of a payment
- What qualifies the contractor for final payment?
- Is early occupancy allowed before final payment?
- What percentage will be retained, and for how long?
- What interest rate applies on delayed payments?

Liens and waivers:
- Will lien waivers be required in exchange for payment?
- Can the owner deduct from the amount due for lien claims?

Insurance:
- What coverage is required?
- What proof of coverage is required?
- Are others insured?

Optional Contract Terms

Depending on the job, your contracts may need to cover an additional 18 subjects. Scan down this list and you'll see dozens of issues you may want to cover in trade contracts. Each of these topics has the potential to become a dispute. Settle these issues in the contract before the work starts.

Site safety, protection and emergencies:
- Who is responsible for project safety?
- Who is responsible for fire safety?
- What response is required in an emergency?
- Who is responsible for protecting new and existing work?
- Who supplies fencing and toilets?

Hazardous materials:
- Any limits on use of asbestos, flammables or lead?
- Will explosives or welding equipment be allowed on the site?
- Who is responsible for hazardous materials discovered on-site?

Survey and layout:
- Who will do the survey and job layout?
- Who is responsible if there are errors in the survey?

Permits, fees and taxes:
- Who applies for the permit and gets approvals?
- Who pays for permits and approvals?

Utilities, cleanup and job signs:
- Who provides temporary utilities?
- Who arranges for permanent utilities?
- What are the owner's rights if the jobsite isn't kept clean?
- Is a job sign either allowed or required?

Role of the superintendent, architect or engineer:
- What is the authority of the owner's representative?

Subcontracts and subcontractors:
- Can the owner reject subcontractors?
- What obligation does the owner have to subs?

Use of the site and adjacent property:
- Any restrictions on use of the jobsite by the contractor?
- Does the owner have free access to the site?

Responsibility for surprises in the job:
- Who is responsible for a mistake in the plans?
- Will the contractor get extra compensation for surprises?

Changes in the work:
- ➤ Can the owner insist on changes at a set cost?
- ➤ Who pays for changes required by law or a plan defect?
- ➤ Is there a formula for pricing extra work?
- ➤ Are losses charged to the contractor?

Warranties, defective work and callbacks:
- ➤ Can the owner reject work considered defective?
- ➤ What defects qualify for a callback, and for how long?
- ➤ What warranty is included?

Handling claims and resolving disputes:
- ➤ How will disputes be settled?
- ➤ Is a written notice of claim required?
- ➤ Is arbitration required?

Liability for damage, indemnity and bonds:
- ➤ Who is liable for damage to the work?
- ➤ Will the contractor be liable for accidental losses?
- ➤ Will performance and payment bonds be required?
- ➤ What are the limits to liability?

Contract boilerplate:
- ➤ Can rights under this contract be assigned to others?
- ➤ Do waivers have to be in writing?
- ➤ What are the rights of third parties?
- ➤ Does the owner have the right to audit records?
- ➤ Are there any limits to contract claims?

Inspections, testing and delays:
- ➤ Who schedules tests and inspections?
- ➤ Who pays for inspections and re-inspections?
- ➤ Is the contractor required to uncover work for inspection?
- ➤ What has to be in the written schedule?
- ➤ What type of contractor delay will be excused?
- ➤ Will there be liquidated damages for a nonexcusable delay?
- ➤ Can the contractor collect for delay by the owner?
- ➤ How is the cost of delay figured?

Suspension and termination of the job:
- ➤ Does the owner have the right to terminate the job?
- ➤ Does the contractor have the right to stop work for nonpayment?

Completion:
- When is the job substantially complete?
- Will there be a punch list?

Contract Bias

"The bottom part is our standard disclaimer. It says we're not responsible for anything."

All contracts are not created equal. Most attorneys will confirm that they could write a construction contract that either:

1. makes it nearly impossible to lose money on the job, or
2. makes losing money on the job almost certain.

It all depends on contract bias. Every contract has bias, favoring either the property owner or the contractor. You have better control, more options and extra security when contract bias is in your favor. Don't believe any claim that a contract is "standard" or has no bias. There is no such thing as a standard construction contract, just as there is no such thing as a standard construction project.

Figure 6-1 illustrates contract clauses that shift bias to favor either the contractor or the property owner.

> *"Every contract has bias, favoring either the property owner or the contractor."*

Using the contract clauses listed, it's easy to create a very one-sided contract. But even if you could, there's no need to bend every contract clause your way. A more balanced contract will be accepted sooner and with fewer revisions. If you want to shift contract bias your way, be selective. For example:

- If the contractor is financially solid with a good reputation for meeting obligations, bias in retainage and release of liens may be irrelevant. There's no need to weight those clauses heavily in favor of the owner.

- If the owner is in a hurry to take occupancy, pay close attention to bias when dealing with the construction schedule.

- Suppose you think competing contractors may plan to bid the job at cost and then make a mint on change orders. Consider bias very carefully when selecting clauses that spell out charges for extra work.

Contract Terms

That Favor the Contractor

- The contractor's estimate defines the scope of work.
- The contractor can rely on any claim made by the owner.
- Design defects are corrected at the owner's expense.
- No retainage will be deducted from payments.
- The owner can't make changes without consent of the contractor.
- Changes required by law are charged as extra work.
- An error by the owner may result in extra charges.
- Changes will be charged at the normal selling price of contractor.
- The owner pays if labor or material costs increase after the contract is signed.
- The callback period ends at final completion.
- The contractor provides no warranty other than required by law.
- Disputes have to be resolved by arbitration, not litigation.
- The contractor is compensated for a delay caused by the owner.
- The contractor can suspend work or terminate the job for slow payment.
- On termination, the contractor recovers the full contract price, less the cost of completion.
- Final payment is due 30 days after termination.
- Any defect not on the punch list is accepted.
- The prime contractor isn't required to pay subcontractors until paid by the owner.

That Favor the Property Owner

- The job includes everything that can be inferred from the plans and specs.
- The contract price covers job conditions as they exist.
- The owner disclaims any assumptions made by the contractor.
- The owner can discharge employees or terminate subcontractors for cause.
- The owner isn't responsible for jobsite safety.
- The contractor is responsible for hazmat found on the job.
- The owner can reject any work considered defective.
- Any ambiguity in the plans is resolved in favor of the owner.
- Final payment releases all contractor claims.
- Retainage will be deducted from all payments, including change orders.
- The owner can withhold payment for failure to follow the plans.
- The owner can withhold payment for failure to keep the work on schedule.
- The owner can require that changes be made at the contractor's cost.
- Claims for extra work expire if not made in writing within five days.
- Unsatisfactory work must be removed at contractor's expense.
- The contractor provides a broad form warranty on all work completed.
- The contractor is liable for any loss suffered by the owner (indemnity).
- The contractor has to pay for any corrections required by the inspector.

Figure 6-1
Contract bias

Comply with State Law

Every state has the right to legislate what has to be in a construction contract — and nearly every state has done so. Since the law is different in every state, construction contracts have to be different in every state. No construction contract complies with the laws in all states.

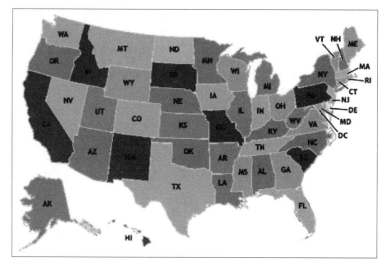

Select your state at http://Construction-Contract.net

Penalties for Using an Illegal Construction Contract

Penalties for using a contract that doesn't comply with state law can be severe. Courts in most states won't enforce an illegal contract, often leaving a contractor with no way to collect. Where contractors are licensed or registered, use of an illegal contract can result in discipline (fine or suspension). Penalties vary by state. To find out what they are in your state, browse to the website www.Construction-Contract.net. Then click on your state to see a list of penalties for using a construction contract that doesn't comply with state law.

Every state has special requirements for construction contracts. In most states, courts won't enforce a contract that omits a specific notice or disclosure. What's required by the law in your state depends on:

> **The type of construction** — what's getting built. For example, work on pools, insulation, roofing, basements, home improvements, landscaping, electrical and mechanical work require specific contract provisions in many states.

> **Who signs the agreement.** Prime contracts (between a contractor and a property owner) have to follow one set of rules. Subcontracts (between two contractors) follow different rules.

> **The value of the work.** Contracts for work valued below certain minimums (such as $500) are exempt from some state requirements. Contracts for larger jobs (such as over $1 million) have other requirements.

> **Residential or non-residential.** Residential contracts for work in states with the largest urban populations (CA, DC, FL, IL, LA, MA, NJ, NY, OR, PA, TX) require the most notices and disclosures. For example, California requires 18 distinct notices and disclosures for some types of residential work.

> **Where the contract is signed.** In many states, a contract signed or negotiated at the home of one party to the contract will be void and unenforceable without specific disclosures. If a construction contract is void under state law, mechanic's lien rights may be forfeit.

Contracts for commercial work don't require as many notices and disclosures as contracts for residential jobs. But many states have prompt payment acts which apply on commercial jobs, especially larger commercial jobs (over $1,000,000). Most prompt payment statutes require that contractors and subcontractors be paid on a specific schedule. State law usually voids any payment term in a contract that doesn't comply with the statute — and may impose a penalty for failure to pay on the state's schedule.

Construction contracts also have to comply with federal law. The best known of these federal laws mandates a Three-Day Right of Rescission. When work is done on the principal residence of an owner, multiple copies of the three-day notice have to be delivered with the contract. If the contractor is extending credit or arranging financing or if payments will be made after work is completed, the Federal Truth-in-Lending Act requires additional disclosures. Federal law applies in all states, regardless of what state law may also require.

Staying Legal

Obviously, you want your trade contracts to comply with state law — you don't want to distribute an illegal construction contract as part of any bid package. But how can you be sure the contract you're using complies with state law?

Many model construction contracts are available. Unfortunately, none of the model contracts provided by the following associations are specifically intended to comply with either state law *or* federal law:

> The American Institute of Architects (AIA) contracts — probably the best known

> The Engineers Joint Contract Documents Committee (EJCDC)

> Design-Build Institute of America (DBIA)

> Consensus DOCS, sponsored by several construction industry associations

Contracts offered by the AIA tend to protect the architect or designer. Contracts offered by the EJCDC protect the engineer. Contracts offered by the DBIA protect the property owner. Consensus DOCS are *supposed* to have a neutral bias, not favoring the architect, engineer, owner or (presumably) the contractor.

So where do you get legal construction contracts? That's not an easy question to answer. We don't know of any state that publishes a concise list of all the notices and disclosures required by state law. Two states, Oregon and Maine, suggest contract clauses drafted to comply with state law. But both of those states caution that the samples offered shouldn't be taken as authoritative. And none of the samples published by Oregon or Maine meet federal requirements. In short, states set stiff standards for construction contracts but do very little to help contractors meet those standards. Basically, you're on your own.

Your Own Contract

Many sample contracts are for sale on the Web. But be careful. We don't recommend buying any sample construction contract that's guaranteed "legal in all 50 states." There's no such thing, for reasons explained in the preceding paragraphs. Nor do we recommend buying any state-specific sample contract unless the variables are clearly defined:

- type of construction
- value of the work
- residential or commercial
- where the contract is signed
- contract or subcontract

As we explained at the beginning of this chapter, most experienced construction contractors have a model contract they prefer — often developed with the aid of an attorney. If you want to develop your own contract, start with printed model contracts offered by a construction contract vendor. A search for "sample construction contract" on the Web will turn up dozens of vendor sites. In most states, you'll need a model contract for each type of work you handle (residential or non-residential, custom home or home improvement). You'll need a contract for larger jobs and smaller jobs, and for work when the contract is signed at some location other than your office. Type those sample contracts into your favorite word-processing program, leaving blanks where terms and prices have to be filled in.

When done:

- Check your model contract against the contracts at http://Construction-Contract.net.
- Have an attorney review your model contracts. Select an attorney familiar with design and construction work and contract requirements in the state or states where you plan to work.

No model contract is going to fit all of your projects. The contract you prepare has to be appropriate for the size of the project you're doing and the client that you're representing. If you're doing a $10,000,000 project for a big car dealer, by all means, start negotiations with your 60-page killer contract. If you're doing a $300,000 project for Mr. and Mrs. Homeowner, the legal requirements are entirely different and a 60-page contract would be overkill.

Boil that 60-page contract down to about a dozen pages that include just the essentials. Then add all the notices and disclaimers required in your state for residential work. If you drop a 60-page contract on homeowners, you'll watch the blood drain from their faces, and they'll promptly lead you to the door and say goodnight. But what they're really saying is goodbye.

Ordering Materials and Services

You'll remember the sad case of Thurber Lumber Co. Inc. v. Marcario back in Chapter 2. The contractor, Joe Marcario, ordered materials from Thurber Lumber Co. for the Nenninger job. Thurber Lumber and Joe Marcario had been doing business for years. But the Nenninger job was different. This time, Joe intended to handle the Nenninger job as a CM contractor. Nenninger would be liable for the cost of materials.

There's nothing wrong with that. It's done routinely on public works projects. A construction manager, working for the public agency as a consultant, orders all materials, probably on a form supplied by the public agency. Suppliers issue an invoice to the public agency, not the construction manager. There's no question about who's liable. The construction manager may get a phone call if the public agency doesn't pay when expected. But no court would hold a construction manager liable for materials ordered on a public agency form.

We recommend following the same procedure any time you work as a construction manager (consultant). It's perfectly acceptable to order on behalf of the property owner, either over the phone, by fax or email. But leave no doubt about who's liable for payment. The Purchase Order Cover Sheet, shown in Figure 6-2, should eliminate any possibility of confusion about who will be liable for payment. Attaching the Purchase Order Cover Sheet to purchase orders would have saved Joe Marcario from a $59,391.51 judgment.

Requests for a Quotation

If you plan to take competitive bids for materials and services, be sure bidders understand who is liable for payment. Use the Request for Quotation Cover Sheet (Figure 6-3) to eliminate any possibility of confusion.

These forms are only cover sheets. Your purchase order or request for quotation has to identify at least the following seven items:

1. What's needed (leaving no ambiguity)?
2. What quantity is required?
3. When is it needed?
4. Who pays for delivery?
5. What's the cost?

Purchase Order Cover Sheet

Details about this order appear on the following page(s). This purchase order is for the job identified in Box 2. Direct any questions about this order to the Construction Manager in Box 4.

Date of this Order (Box 1)

_____
Date this order is submitted

_____
Order identification number

The Project (Box 2)

_____
General description

_____
Location

The Project Owner (Box 3)
This order is placed by the Project Owner, who is liable for payment. Be sure all invoices and shipping documents reflect the name of the Project Owner. Invoices and shipping documents that show any other party will be returned.

_____
Name

_____
Address

_____
Address

_____
Fax number

_____
Voice phone

_____
Email address

The Construction Manager (Box 4)

_____
Name

_____
Address

_____
Address

_____
Fax number

_____
Voice phone

_____
Email address

Figure 6-2
Purchase Order Cover Sheet

Request for Quotation Cover Sheet

Details about this request appear on the following page(s). This quotation is for the job identified in Box 2. Please respond to this request by quoting a price and terms. Your quotation should be addressed to the Project Owner. Payment will be made by the Project Owner. A quotation submitted to any other party will be declined. Forward your quotation to the Construction Manager in Box 4.

Date of this Request (Box 1)

Date this request is submitted

Project I.D. number

Who's requesting the quote

The Project (Box 2)

General description

Location

The Project Owner (Box 3) Respond to this request by quoting price and terms for payment by the Project Owner. Be sure to note any discount for prompt payment.

Name

Address

Address

Fax number

Voice phone

Email address

The Construction Manager (Box 4)

Name

Address

Address

Fax number

Voice phone

Email address

Figure 6-3
Request for Quotation Cover Sheet

6. Any additional charges that may apply
7. Any discounts or restocking charges?

Be sure the quotation matches what's required for the job. For example, the specs may require testing or inspection before installation is considered complete. Who will do that testing or inspection and who pays? What happens if the installation doesn't pass inspection? Is any certificate required? Will shop drawings or samples have to be approved before installation? Does the quotation include any warranty required by the contract documents? If the project is intended for LEED certification, many materials will require a certification by the supplier.

Concrete, steel welding and bolt pull-out tests are the most common tests required by engineers and building departments. A concrete mix design may have to be submitted to the structural engineer for review before ordering concrete.

Construction specifications often reference state, national or international standards, such as the following:

> ACI — The American Concrete Institute's Manual of Concrete Practice

> AISC — American Institute of Steel Construction's Manual of Steel Construction

> ANSI — American National Standards Institute

> ASHRAE — The American Society of Heating, Refrigerating, and Air Conditioning Engineers

> ASTM — Originally known as the American Society for Testing and Materials

> AITC — The American Institute of Timber Construction's Timber Construction Manual

> NFPA — The National Fire Protection Association's *National Electrical Code*®

The supplier should be prepared to provide documentation verifying compliance with any manufacturing standard cited in the job specs. If a standard requires a specific installation procedure or measurement of performance after installation, either the trade contractor or the supplier will have to arrange for inspection. If inspection and testing aren't part of the trade contract, include inspection or testing in your purchase order or request for quotation.

Get Started Building

If the plans and specs have been approved and the contracts have been signed, it's finally time to start work. That's what we're going to do in Chapter 7.

Chapter 7

Observing Day-to-Day Construction

EVERY PROJECT REQUIRES SKILL, ingenuity and attention to detail. If you've been working in construction for more than a year or two, you have a pretty good idea about what's required to complete a project. Almost certainly you know that:

- ➢ construction is complex
- ➢ there's no single best way to handle any job
- ➢ it's easy to make a mistake
- ➢ construction is too permanent and too expensive for trial and error
- ➢ there's no substitute for spending actual time on the jobsite

And very likely, you understand the role of a general contractor or construction supervisor. This chapter will concentrate on work unique to CM contracting. That role can be summarized as:

> *Monitoring work on the project both on and off the jobsite to ensure that construction means, methods, techniques, sequences and procedures followed by contractors, suppliers and service providers are in compliance with their contracts and agreements, and that construction is proceeding as planned.*

Notice that what a construction manager does is very similar to what any general contractor or project superintendent does. The difference is in the word *monitoring* in the paragraph in italics on the previous page. A general contractor or job superintendent doesn't just monitor the work. A general contractor or job super *directs* the work, giving orders, making corrections, resolving problems on the spot. A prime contractor is the final authority. The same can't be said of a CM contractor. The owner is the final authority. A CM contractor is simply a consulting advisor to the owner.

"Direct, Oversee, Monitor and Advise"

Each of these terms defines a specific responsibility. And each term carries important distinctions you should understand. I'll use my (Bill Mitchell) background to provide a few practical examples.

During my 50 years in construction, I've served in many roles. Each was intended to meet requirements of the owner. On some projects, I was simply the architect. In that capacity, I did the design, got the permit issued and solicited bids for the work. When work started, I was, by contract, required *to observe the work* and report back to the owner weekly. The general contractor *directed the work*. It wasn't my business to tell the subs what they had to do or had done wrong. My job was to make sure the contractor followed the plans. If he didn't, I brought it to his attention. If that didn't work, I'd whine to the owner, who would then tell the contractor to comply or he wouldn't get paid.

On jobs where I had a CM contract with an owner, I usually wasn't the architect. Instead I was the owner's representative. On those jobs, I didn't have the rights and obligations of an architect. Rather, I served as the eyes and ears of the owner. My role was to monitor the work and advise the owner. I neither directed the work nor observed the work as an architect would. I wasn't the general contractor or the architect of record.

There were times when I acted in the capacity of program manager. This role carries a much broader set of responsibilities. I directed everything on very large, very complex projects. As the program manager, I had responsibility for complete *project oversight* of everyone (except the owner) and everything on the project, from beginning to end.

I spent five years as program manager for a school district. We had 13 construction sites and a budget approaching $100,000,000. We had contracts with three design firms (one of whom I had to fire) and two general contractors. Here again, I was *advisor to the owner*. My responsibility was to *monitor* the entire design and construction process. I directed all of the activities and all the players. But I had no right to direct any subcontractor. I could advise and make recommendations to the owner. If approved, I could implement those decisions.

Any time I acted as CM contractor, I was the *owner's advisor* and *monitor*. I had no right to direct actual construction. That was the responsibility of the prime contractor, and ultimately, the owner.

Ask nearly anyone who makes a living as a CM contractor. The distinction we've just made between the role of a general contractor and the role of a CM contractor is more apparent than real. True, the owner is the final authority under a CM contract. The CM contractor only advises. But failure to take a CM contractor's "advice" can have serious consequences for any trade contractor or supplier. Only rarely will a threat of consequences be required to gain compliance. An effective CM contractor finds ways to get cooperation without intervention by the owner or the A/E team.

This chapter will describe a CM contractor's responsibilities for day-to-day operations, starting with threshold issues (bonding and insurance) and continuing through final cleanup. Project closeout will be the subject in Chapter 13. We'll dip into this subject — the role of a CM contractor — once again in Chapter 10. That's where you'll find a concise list of a CM contractor's authority.

Bonds and Insurance

If you've worked on larger jobs, you've probably had to provide certificates of insurance and either bid, performance, or payment bonds. As a CM contractor, your responsibility is to review bond forms and insurance certificates supplied by others. You can do this review any time after the contract is signed. The only requirement is that bonds and insurance certificates be delivered to the owner before materials, equipment and crews arrive on-site.

The construction contract probably sets standards for insurance — types of coverage and the policy limits. Typically, these policies are written with limits of $1,000,000, $2,000,000 or $3,000,000. The contractor's general liability/ fire insurance policy will usually cover travel, cars and equipment used by construction personnel. Even if the contract doesn't mention insurance, you'll want to resolve insurance and bonding issues before work starts. If bonds are required, you should have on file a certificate of coverage issued by the bonding company. The owner's coverage is important too. The contractor may want to see a certificate of insurance showing that the owner has coverage on completed work. If the work is damaged during construction, the contractor's insurance may not cover the loss. Instead, the owner's builder's risk insurance should provide coverage. It's best to have an insurance agent or broker compare the owner's and contractor's policies to ensure that all significant risks are covered and that there's no double coverage.

If the contract identifies the coverage required and policy limits, contractors should submit the draft contract to their insurance carriers and bond underwriters before bidding the job. That will eliminate most surprises. Most owners will require that they be named as an additional insured on the contractor's policy. That usually increases the cost of coverage. Some carriers refuse to write insurance for certain types of risks, while others restrict policy limits. All bond underwriters set limits on the bonding capacity any contractor can carry. Contractors should satisfy themselves that the required coverage is available and understand the cost of that coverage before bidding the job. Note that Box 7 of the *Invitation to Bid* (in Chapter 5) includes an option to require bonds and certificates of insurance as part of the bidding process.

Insurance Coverage

There's no need to see a copy of a contractor's insurance policy. Instead, request a certificate of insurance. Insurance agents and brokers issue these routinely upon request, usually at no charge. The certificate confirms that insurance is in effect and will identify details about the insurance coverage — name of the insurance company and a phone number, the name and address of the insured, type of coverage and significant exclusions, policy limits, policy number and effective dates. Request a certificate for each policy that should be in force. If the contractor will have subcontractors working on the job, request a certificate of insurance from those subcontractors.

Be sure that the policy will cover the full project duration. A policy issued on an annual basis should be renewable. Performance and payment bonds should be renewable for the anticipated project duration. A contractor's bonding capacity can change from month to month and will vary with the volume of work in progress. Limits in bonding capacity can make it difficult to protect a project which runs for several years.

Most carriers issue certificates of insurance on forms developed by ACORD Corporation. The standard ACORD form commits insurance carriers to "endeavor" to give notice of cancellation, but disclaims liability for failure to give that notice. If the owner insists on notification 30 days before cancellation of insurance, a nonstandard certificate of insurance may be required.

Types of Coverage

Comprehensive General Liability (CGL) — This covers claims made against a contractor based on negligence, either of the contractor or a subcontractor or of a third party under an indemnity agreement. Completed operations coverage (product liability) may also be included. The owner and the CM contractor may be named as additional insureds under a CGL policy. If the insurance carrier won't add the owner and CM contractor as additional insureds on a single policy, an OCP (owner and contractor's protective liability) policy makes a good substitute. An OCP policy can protect the owner, engineer, architect or CM contractor against liability of a specific contractor on a specific construction site.

X C U Liability Coverage — Most CGL policies exclude coverage for explosion (X), collapse (C) and underground hazards (U). If the work includes blasting, high-pressure piping, or power-transmission equipment, explosion coverage should be required. Collapse (C) coverage provides insurance for structural injury to a building that results from excavating, tunneling, pile driving, shoring or demolition. Underground (U) liability insurance covers losses to underground utility lines caused by excavation equipment.

Worker's Compensation Insurance — Nearly all states require worker's compensation (or equivalent) coverage for employees. In states where contractors are licensed or required to register, the contractor may have to show a certificate of coverage as part of the application process. Worker's comp insurance provides payment for medical care and disability benefits under a schedule established by law. Both prime contractors and subcontractors need coverage for all their employees. If the job qualifies as an owner/builder project under the building code, the building department will probably require the owner to show evidence of worker's comp coverage before pulling the permits.

Builder's Risk Insurance — Covers property damage to the project while under construction. This coverage is usually provided by the property owner. That makes sense because work completed is usually considered to belong to the owner. A stack of lumber delivered to a jobsite belongs to the contractor and should be insured by the contractor. When that same lumber is assembled as framing, it belongs to the owner and should be insured by the owner.

Vehicle coverage — Should include liability and property damage coverage for owned, hired and non-owned autos.

Some large projects have a different type of coverage called *Wrap Insurance* or *Owner-Controlled Wrap-up Insurance* (usually called a *wrap-up policy*). This type of coverage is purchased by the owner or developer and covers nearly all insurable risks for the owner and all contractors and subcontractors on the project. The wrap-up program is usually administered by an insurance agent or broker selected by the owner or developer. The cost of coverage is charged proportionally to each contractor and subcontractor working on the job. Wrap-up insurance is controversial because contractors and subcontractors (1) lose the right to negotiate their own coverage, (2) get charged at rates they may not have approved in advance, (3) have little control over settlement after a loss, and (4) are bound by decisions of others on policy limits and loss deductibles. Wrap-up insurance works best on larger projects (over $25 million) and on residential tracts because it eliminates disputes about which contractor or subcontractor was at fault. All are covered by the same carrier. That tends to reduce the owner's cost of insurance.

Sometimes you'll see construction contracts that set high policy limits for various

---

An Insurance Horror Story

Don't think of insurance as an expensive nuisance. Insurance is as essential to construction as tool belts and hardhats. I'll tell a short story to illustrate the point.

I few years ago, I drew plans for a three-story medical office building on top of a hill in Escondido, California. We'd pulled the permits and poured the foundation. Then the framing sub started work. This particular framer was out to impress the owner with his speed and efficiency. He and his crew literally threw up the entire three stories in just *two weeks*. One reason the frame went up so fast: The framer didn't install plywood shear wall panels and floor-to-floor steel ties.

On the last day of this framing race, the whole crew went out to lunch. During that lunch break, a freak wind gust came up the mountain and hit the unbraced framing. The entire 20,000 square feet of framing twisted sideways and rotated into the parking lot. Fortunately, no one was hurt, as no one was there.

The owner was on the phone with me in 15 minutes. He was more than a little concerned. Remember, once framing is up, it's the property of the owner. The next morning I was on-site taking pictures and writing up juicy notes. Remember too, that an architect's plans show how a building is to be when it's finished. The way the building gets into that final condition is at the sole discretion of the contractor. The contractor's process is known as the *means and methods* of construction. In this case, the plans were fine. It was the means and methods that were faulty. The framing sub's insurance carrier covered the cost of pulling all the lumber apart and reassembling it as shown on the plans. But this time, shear walls and steel straps went up as each floor was finished.

types of insurance and require that all contractors and subs carry similar policies. Owners seem to believe that if some insurance is good, more has to be better. Duplicate coverage is common when multiple design and construction firms are involved on the same site. There's no advantage to double coverage. It's simply a waste of money. Claims are going to be paid only once, no matter how many times the loss is insured. Better to have an agent or broker tailor the coverage of all concerned to eliminate double coverage and plug any coverage gaps.

Surety Bonds

Bonds are more common on public works projects than on private jobs. On private contracts, suppliers and subcontractors have mechanic's lien rights. If the prime contractor defaults, subs and suppliers can file liens against the property itself. To clear title to the property, the owner will have to discharge the liens. Eventually lienholders who comply precisely with the lien law will be paid.

There's no such guarantee on public works projects. Only a few states allow liens on government property. Instead, state law or the federal Miller Act requires that the contractor buy a bond (payment guarantee) issued by a surety company licensed to do business in the state. That's a *payment* bond.

Here are some bonds commonly used in construction:

> A *bid bond* is a guarantee by a surety company that the contractor will sign the contract if he's selected to do the work.

> A *performance bond* guarantees completion of the project once the contract is signed. Bonding companies usually issue bid bonds and performance bonds in tandem.

> A *maintenance bond* is a guarantee that work completed will meet specific performance characteristics for a specific period.

> A *permit bond* may be required when work has to be done on public property, such as when connecting a sewer line in the street. The bond will be released when the street is restored to proper condition.

> A *defect bond* can be issued when a dispute over some item is delaying final payment. The owner may be willing to release final payment if the contractor *bonds off* the disputed item — buying a bond equal to perhaps 120 percent of the estimated cost of resolving the claimed defect.

On public works jobs, a subcontractor who has a dispute with a general contractor can file a stcp notice with the owner. The law in many states then requires the owner to withhold 125 percent of the amount of the claim until the dispute is resolved. To collect from the owner, the contractor either has to pay the sub or buy a bond that guarantees payment of the sub.

A contractor who supplies a bond usually negotiates terms with the bonding company and will pay the bond premium. All bonds reduce the risk of loss to the owner. All bonds also increase the cost of completion. Ultimately, the owner pays the cost of the bond in the form of a higher construction cost. That's why bonds are less common on private construction projects. The cost of a performance bond is usually from 1 to 2 percent of the contract price, depending on the surety company's evaluation of the risk. If the contractor defaults under a performance bond, the bonding company will usually hire another contractor to finish the work and then try to recover against the defaulting contractor.

Default insurance is an alternative to performance and payment bonds. Like a performance bond, default insurance covers the cost of completion after default by a contractor or a subcontractor. Unlike a bond, default insurance covers a broad range of additional costs associated with a default, such as legal fees, higher overhead and the cost of delay. Default insurance also guarantees payment of bills for materials, labor and equipment. The cost of default insurance is usually about the same as the cost of a bond, but deductibles and co-insurance are higher. Most default insurance is written to protect contractors against default by their subcontractors. Not all insurance carriers offer default insurance.

If a bond is required, the contractor should supply a bond form countersigned by an authorized agent of a bonding company licensed to do business in the state. Check to be sure that the full name and address of the property owner appears on the bond and that the amount of the bond is correct. Then pass the bond to the property owner with your recommendation to accept or reject the bond supplied.

Staying on Schedule

One of the most important responsibilities of any CM contractor is to keep the job moving. Every CM contractor has an obligation to avoid wasted time, and to recommend mid-course corrections when needed. Your contract with the owner probably includes a pledge such as:

> *If any phase of construction is not proceeding as planned in spite of the best efforts by CM contractor, CM contractor will recommend steps the owner should take to ensure successful completion of the project on time and within budget.*

Back in Chapter 4, we recommended creating a schedule appropriate for the job. That could be as simple as a few notes on a calendar, or as complex as you want to make it using one of the popular software tools such as *Microsoft Project*. In Chapter 4, no contractor had been selected. So any schedule for the job was almost certainly your responsibility. From the time work starts, at least one trade contractor will be involved in planning and scheduling the job. In most cases, the prime trade contractor will develop a detailed written schedule for your approval. Once work starts, you'll measure progress against that schedule and require revisions when progress doesn't match the approved schedule.

Your responsibility as a CM contractor is to ensure that:

- schedules prepared by trade contractors are appropriate for the work at hand
- work is proceeding in compliance with the approved schedule
- remedial steps are taken if the schedule starts to slip
- schedules and delivery times harmonize to minimize conflicts

"You schedule 28 crews on 18 different jobs, remember the date of over 300 progress payments — but you can't remember where you put your car keys?"

A good construction schedule makes it easy to compare job progress to date with expected progress to date. That comparison is almost automatic when due dates for progress payments are based on completion of job phases. For example, suppose you receive an invoice dated June 10 for foundation work. The contract provides that payment for foundation work is due when the inspector signs off on the foundation work. Checking the schedule, you note that foundation work was to be completed on June 10. It's easy to conclude that the project is on schedule.

If payment due dates are based on the calendar rather than on construction phases, monitoring the schedule becomes a separate task. We'll have more to say about the relationship between scheduling and invoicing in Chapter 9.

What Should Be on the Schedule?

If the contract between the owner and the prime contractor requires a construction schedule, the same contract probably requires approval of that schedule before the first payment is due. Responsibility for reviewing the proposed schedule falls on the CM contractor. So, what should you look for when reviewing the construction schedule?

The first issue will be contract compliance. The contract could require a bar chart schedule, a CPM (critical path method) network diagram, a narrative schedule or even use of a specific scheduling tool (such as Primavera *SureTrak*).

A detailed schedule will show the beginning date and completion date of every significant task in the job. Obviously, what constitutes a "significant task" is subject to interpretation. For a larger, more complex job, the schedule should show every distinct activity expected to take one day or longer. Any activity expected to last more than two weeks should be broken into two or more activities.

A Scheduling Horror Story

About three years ago, I was asked by a school district to oversee the re-roofing of several free-standing buildings on several separate school sites. They didn't have anyone on staff who could manage such a large set of roofing projects. So I agreed to serve as the CM contractor. Here's what they *didn't* explain:

➢ the mandatory job walk was scheduled for the next week

➢ the bid date was in less than 30 days

➢ plans and specs hadn't been drawn yet

This job was going to be a scheduling nightmare.

I cancelled the job walk and the bid opening. Next, I hired an outside architect who rummaged through old plan sets and found enough site, plan and detail stuff to cobble together a marginal set of bid documents. We reset the bid date and selected the low bidder. I figured we were back on track.

That's when flaws in the public bidding process kicked into high gear. Our apparent low bidder was based 100 miles away. They were a large firm, as roofing companies go. But they didn't have the slightest clue about job scheduling.

What should have been a 90-day project turned into 12 months of agony. I fought tooth and nail to keep work moving. After nearly a year, they were down to our final building — a large gymnasium with two large rooms, each with a basketball court surrounded by bleachers. It was now late fall. The weather was beginning to change. So, what did these guys do? You guessed it, their demolition crew tore off the existing roof and left town for a month. Now it's November. I can see dark clouds forming over the mountains. I called. I called. I called. I sent faxes. I whined to the owner. It was, after all, their roofing contractor. I was just the CM contractor. Long ago, I advised the district to terminate these guys.

Now it's Wednesday. Dark clouds are right over the city. I called and I called and I finally got the roofer. I told him it's going to rain. "You've got to get the gymnasium roof covered." He agreed. "We'll get right on it," he told me.

Now, it's Friday and still the roof isn't covered. I called. I faxed. I emailed. No response. Then it's Saturday. I'm home, 20 miles away, and it's starting to sprinkle. I call the roofer and get the project manager. I tell him it's starting to rain. He says he has a crew on the roof as we speak, covering it with plastic. "Don't worry. We've got it under control."

At 6 a.m. Monday morning, my cell phone rings. It's the school's head of maintenance. He's standing in a puddle of water covering the entire gymnasium floor. Water is pouring through cracks in the roof. What should he do? I say, "Keep buckets under the leaks and sweep out the water. I'll call the roofer and be there in an hour."

To say it was a mess would be an under-assessment. The plastic had blown off the roof. The hardwood floor was expanding, like the sides of a wood rowboat on the way to Catalina. Three hours later, the roofing project manager arrived — to explain how bad it *wasn't*. With a couple of hair dryers and a little buffing, the floor would be back to normal in a couple of days. Right? Well, not exactly.

(Continued)

> **A Scheduling Horror Story,** *continued*
>
> I met with the district superintendent. We called the school's insurance carrier who immediately took over the job from the roofer. Two and a half months, 400 photographs, 16 enormous fans, four sanding machines, floor striping and finishing painters, new wall and roof insulation and a new roof later, the kids moved into a nearly-new gym. The insurance cost? Somewhere north of $600,000.
>
> Moral to this story: Don't deal with a contractor who can't create a schedule and commit to that schedule. A low bidder who can't stay on schedule is no bargain. If we had hand-selected a competent local roofing sub, this CM contractor would have been back home in 90 days, counting my money.

A detailed schedule will include:

- a description of the activity
- the duration, including dates for early or late start and early or late finish
- where the activity falls on the critical path. Any delay on the critical path holds up the entire job
- a completion date that is the sum of all tasks on the critical path
- total *float* (idle time) for each task. Use float as an allowance for bad weather
- dates for inspections and tests
- start dates and completion dates for subcontractors
- lead times and order dates for key materials
- dates for approval of submittals, shop drawings and samples
- conditions which require a revised schedule
- a legend which defines each abbreviation and symbol

Inspection milestones are a necessary part of every schedule. There's no need to guess about when a project passed a particular inspection. Be sure each inspection and required test is noted on the schedule. For residential work, most jurisdictions will require the following inspections:

- rough and finish grading, foundations and footings
- underground plumbing and electrical
- concrete forming and rebar
- wall and floor framing
- floor and roof sheathing

- exterior insulation, siding and roofing
- rough and finish plumbing
- mechanical heating and cooling
- rough and finish electrical
- elevators, lifts, ramps and conveyors
- fire protection
- drywall, taping and painting
- punch list and close out

Other inspections and materials testing may be required, including inspections by specialists not working for the local building department. Among these will be inspections required by the structural or civil engineers.

For large jobs, keep a log book of submittals, change orders, etc. We'll have more to say about keeping a job log in the next chapter.

When any subcontractor will be on the job for more than a week or two, the schedule for that sub should be prepared from milestones provided by that subcontractor. Major subs should sign off on the schedule submitted for their work. The same is true for any supplier of key materials subject to a long lead time.

Be sure contractors identify and order long-lead-time items during the first week of the project. The same goes for utility permits. I just finished testifying as an expert witness on a contractor's failure to plan ahead for long-lead-time items. I made good money testifying — probably more than the contractor did on the entire job.

The schedule should include an allowance for both typical weather delays and indecision by public and private authorities. Anticipate delays from private organizations (the utility company, a plan review committee) as well as any government authority (fire marshal, building department). A schedule that doesn't allow for typical delays isn't realistic and offers a poor defense against hostile cross-examination in a court of law.

Contingency, or float (idle) time in a schedule is owned by both the contractor and the owner. That means both the owner and the contractor can spend float to compensate for delay. For example, an owner who's a few days late in approving a shop drawing can invade float time later in the schedule for ordering or installing that particular material. A submittal returned in less time than allowed by the contract can earn a schedule credit.

Before adding float time, check to see how much float for weather and the permit process is already built into the owner's plans. Before building a schedule for a large project, check what float the owner has allowed in the bid documents.

Submittals

Samples, shop drawings or submittals will be required when some construction details are left to the initiative of the trade contractor. If samples, shop drawings or submittals are required, the contract will probably spell out the routing procedure and days allowed for approval. On most typical jobs, submittals will be required two to four weeks after the contract is signed. Submittals help keep the owner (or owner's representative) informed about decisions made by the contractor. The next chapter covers keeping the owner informed. For now, it's enough to flag the issue.

If samples, shop drawings or submittals are required, the following should appear on the construction schedule:

> a description of the sample, shop drawing or submittal

> the lead time necessary to meet the delivery date

> the date the owner will receive a request for approval

> the date approval must be received to meet schedule requirements

Be alert for unrealistic estimates of the time required to approve submittals, samples or shop drawings. Everyone goes on vacation at least occasionally. There will be times when no one is available to approve submittals. And it's unrealistic to assume that every submittal will be approved on the first try. No owner wants to be pressured into approving something simply because time is short. Avoid having many submittals, samples and shop drawings scheduled for approval at the same time. That puts additional pressure on the approval authority. It's better to start approvals early in the project and process submittals a few at a time.

In any case, keep track of dates and deadlines for submittals. A job log book for each day usually works best. We'll have more to say about your job log book in the next chapter.

Approving the Schedule

Approval of the schedule isn't a review of technical accuracy. It certainly isn't an agreement that every activity and every duration in the schedule is correct. What appears on a schedule is only opinion, the informed opinion of the contractor who prepared the schedule. It's just an estimate. You'll want to correct obvious errors and omissions. But you'll seldom have reason to dispute estimates made in a schedule.

Here's what you're checking for:

> compliance with the contract completion date

> enough detail on what has to happen and when (labor, materials, subs)

- a plan that puts each activity in logical sequence
- reasonable assumptions about forced idle time (weather, etc.)
- clear guidance on milestones the owner has to meet
- a definition of what's on the critical path
- a procedure for making (monthly?) revisions when needed

It's time to get out your "Approved" stamp when you're satisfied that the schedule:

- makes it easy to monitor work progress
- provides easy access to what the owner needs to know
- confirms that the contractor has planned each step in the job
- presents a practical plan for completing work within an acceptable time period

Once a schedule is approved, it should become a public document available to all who have a need to know. Approved changes to the schedule should get similar circulation.

Approval of the construction schedule is your pledge not to interfere with work progress that's consistent with the schedule.

A formal written schedule isn't just eyewash. It's a yardstick for measuring progress — a convenient summary. "The job is X days ahead of (or behind) schedule." That's an objective measure easy to communicate to anyone. But that's only an ancillary benefit. The primary purpose of every schedule is to help the contractor anticipate delays and potential problems. In other words, preparing a detailed schedule helps the contractor think through how work will be done. Understand that point and you understand the importance of construction scheduling.

Tips for Tight Schedules

Many construction projects have to be finished on a tight schedule. For example, public school work is best done when students are on summer vacation. Residential remodeling is easiest when the family is elsewhere. On short duration projects, I use a three- to seven-day mini look-ahead schedule to keep a tight rein on progress. On such jobs, the specs and the contract should compress the submittal and submittal review period. Submittals should be in your hands seven to 10 days after the contract is signed. Design and engineering review should take no more than three or four days.

When time isn't a limiting factor, submittals usually arrive within 30 days after the contract is signed. Design and engineering consultants usually have two weeks to review and return those submittals. If not approved, revision and resubmittal may require another week or two. You can't afford to burn six weeks on a 90-day job. Insist on submittals in seven days. Then get the designers and engineers to a meeting on the jobsite the next day with their approval stamps. Get everything signed off then and there.

When I operate as the architect on a job, I shorten the submittal process by bringing my submittal approval stamp to the contractor's trailer at one of our regular job meetings. After the meeting, I stamp and sign off on everything I can. What I can't approve, I hand-deliver to the engineer and stand there while he or she signs off where required on each submittal.

On a short duration job, don't let the designers and engineers opt for revise and re-submittal on shop drawings and submittals. Instead, require approval of those submittals and shop drawings, either without change or with changes as noted. Never, never, never give up time voluntarily. *Never!*

As a CM contractor, one of your primary tasks is eliminating schedule burn. I've worked on at least 50 tightly-scheduled jobs. Only twice have I missed a completion date. On those jobs, I'm glad to share blame with a public agency that couldn't get its act together.

Here's one example on avoiding schedule burn. At a jobsite meeting I attended, the contractor suggested hanging a small air conditioning unit off a roof parapet wall instead of mounting it on the roof as shown on the plans. It looked easier, quicker and cheaper to me. So why not? The architect's representative was there. He said, as most architects would, that he would have to meet with his structural engineer and would get back to us shortly. To me, *shortly* means a week or two. On a tight job, that's absolutely unacceptable.

So I took out my yellow notepad and drew up the detail freehand. The contractor faxed my drawing to the structural engineer. Then I had the architect call the structural engineer and put him on the speaker phone. I explained the detail and the contractor's reasoning and asked if we could proceed as shown in the fax. The structural engineer agreed. I signed off on the change and the contractor began installing the air conditioning unit that afternoon. Everybody was happy. I told the architect to go back to his office and do an "as built" drawing for use when we closed out the job two months later.

To stay on schedule, keep the builder building. As a CM contractor, it's your job to sweep the jobsite clean of problems that could delay progress.

Another Example

Here's another trick you could use to speed delivery of design information. I carry around in my car several cans of bright red spray paint. Suppose the contractor has a site conflict. For example, the plans show a parking lot curb to be built where there's an existing tree. Or a sidewalk is planned where there's a fire hydrant. If the plans aren't changed, the tree and the fire hydrant have to go. So I change the plans by spray painting the new location on the ground. The contractor gets to keep on building. Next, I draw the change in ink on the contractor's plans. I sign and date the change. That leaves a record of the *owner's change*. Then I fax that change to the designer and request an as-built drawing.

If a revision is needed in the building itself, I draw the revision on-site on a piece of drywall or plywood. Then I have the contractor and the sub transfer that detail to their set of plans. I sign off on the *owner's change* the next day. Again, I have the detail sent to the designer/engineer so everyone is fully informed.

The list of a CM contractor's authority in Chapter 10 includes making minor changes — especially changes that don't affect cost, function or schedule. I admit, changes like this can make the designer a little grumpy. The owner's representative is wading around in their swamp! When I get a phone call from a designer who's miffed, I explain that I've just done him a big favor, making a revision that covers his mistake. "If I hadn't done that, the contractor would be on the phone now complaining to the owner. Besides, do you really want to stop what you're doing in the office right now and drive 30 miles down here to argue about a tree or fire hydrant? Because if you do, I'm going to want you here in the next two hours."

Almost without exception, the designer agrees to prepare an as-built drawing that covers the "mistake." To seal the deal, I might buy lunch or offer a bottle of wine or a ticket to the basketball game. But the project keeps moving on schedule.

I'll admit that I have an advantage when working with designers. I have an architect's license, the same as they do. If push comes to shove, I could replace them. But threatening isn't the point. Keeping the project on schedule is always the point. Everybody should understand that from the outset.

Here's a point to remember when dealing with designers. They're far more interested in how a building will look when it's done than how the job gets bolted together. You'll catch more flack from a designer for changing a wall finish or a paint color than you ever will for changing what's behind the wall covering, assuming of course it's structurally sound.

When the Schedule Slips

Many construction contracts list legitimate reasons for delay — reasons for which the completion date is extended automatically without approval of the owner. Many of these reasons fall under the heading *force majeure* (French for *superior force*), sometimes called *Acts of God*. Fire, flood, earthquake, tornado, tsunami, lightning, epidemic and unusually adverse weather are legitimate reasons to extend the completion date without any penalty to the contractor.

Delayed completion may also be excused for labor disputes (strike or boycott), embargo, terrorism, armed rebellion, quarantine, discovery of archaeological or paleontological artifacts, acts or neglect of a public utility or order by a government authority. Unless the contract says otherwise, it's safe to assume that events such as these are legitimate reasons for delayed completion.

If the delay is due to *force majeure,* you'll still want to execute a change order identifying the reason for delay, the dates when the delay started and ended, and the new completion date. That's just good bookkeeping, serving as a reminder for everyone concerned. If there are many delays, documenting each delay as it occurs could prevent a dispute later.

Some contracts provide other valid reasons for delay. These include shortages in labor, materials or equipment beyond the contractor's control. You may want to recommend an extension of time if you get (1) prompt notification from the contractor of the anticipated shortage, (2) convincing evidence that the shortage was unavoidable, and (3) a firm date when the material, equipment or labor will be available.

"The only thing I can figure is that Farnsworth Construction actually finished a job on time."

Don't grant any extension of time without consulting the owner. Every extension of time requires a written change order signed by the property owner. Before agreeing to an extension of time, check for any *float* (idle) time in the schedule. Unless the contract says otherwise, float time in a schedule is generally assumed to be available for use by the owner. Expend float in the schedule before extending the completion date.

If the delay *isn't* due to *force majeure* or some other legitimate reason, the revised schedule should meet the original completion date. The construction contract may require that the contractor prepare a plan to get back on schedule and submit that plan for approval. Many contracts require that schedule revisions be done at no cost to the owner.

There are always good ways to get back on schedule. But nearly all of these will increase the trade contractor's cost:

1. increase crew sizes or the number of crews
2. increase the number of working hours per shift, shifts per day or working days per week
3. reschedule crews so tasks are completed at the same time

Contractor Claims for Delay

Every CM contractor needs to understand the risk associated with construction delay. Construction contractors have a legitimate claim for delay damages when an owner delays the construction process. A recent example may be the best way to make this point. This really happened.

A few years ago, Martin Brothers Contractors won a contract to remodel Crozet Hall, the main dining facility at Virginia Military Institute. Changes requested by VMI during construction delayed the work by 270 days. VMI paid in full for all changes and paid another $99,646.20 for delaying the job. Martin Brothers sued for an additional $330,596.36 in delay damages, including the cost of bringing suit. VMI's response cited two contract clauses. The first allowed damages only for "unreasonable" delay. The second limited delay damages to costs incurred on-site rather than the full cost of delay, which would include home office expense. VMI insisted that the 270-day delay was reasonable and that the claim for an additional $330,596.36 was beyond what the contract allowed. The Virginia Supreme Court (277 Va. 586) sided with Martin Brothers, citing Virginia Code § 2.2-4335, which voids any limit on a contractor's right to recover delay damages. It was an expensive lesson for VMI, nearly $1,600 per day of delay.

The sidebar, on the following page, has more on how courts compute damages for delay.

Any significant delay by an owner can result in an award of damages to the contractor. To reduce the risk of a delay claim, owners prefer to see a "no-claims-for-delay" clause in their construction contracts. For example:

> *No claim for damages or any claim other than for an extension of the contract time shall be made or asserted against owner by reason of any suspension, delay or interruption of the work. Contractor shall not be entitled to an increase in the contract price or payment or compensation of any kind from owner for direct, indirect, consequential, impact or other costs, expenses or damages, including but not limited to costs of acceleration or inefficiency, arising because of suspension, delay, disruption, interference or hindrance of the work by owner or agents of owner.*

If enforced as written, a clause such as this would give the owner an option to halt work at any time and for any period and for any reason (or for no reason at all) and then insist that work start again weeks, months or even years later. Essentially, a true "no-claims-for-delay" clause is a license to bankrupt the contractor. That's why legislatures in many states (including Virginia) have enacted laws which void a no-claims-for-delay clause in contracts for certain types of work.

In most of the remaining states, a no-claims-for-delay clause in a construction contract will be enforced unless the delay:

1. is a type not contemplated by the parties
2. amounts to an abandonment of the contract
3. is caused by bad faith, malicious or grossly negligent acts
4. constitutes active interference by the owner

> **Measuring the Cost of Delay by an Owner**
>
> Contractor claims for delay usually include both direct overhead (jobsite) expense and a proportionate share of unabsorbed indirect (home office) overhead expense. On-site costs tend to be easier to calculate than home office (indirect overhead) costs. A contractor's claim for delay will also include lost profit — usually at 15 percent of all delay expenses. Here are the cost categories you're likely to see when a contractor makes a claim for delay:
>
> *Direct overhead*
>
> > ➤ labor (with taxes, insurance and fringe benefits) for the idle work force
> >
> > ➤ the fair rental cost of idle vehicles, tools and equipment
> >
> > ➤ facilities (temporary structures, water, power, phone, toilets, etc.), sometimes called *general conditions*
> >
> > ➤ the additional cost of bonds and insurance
> >
> > ➤ direct overhead costs of all subcontractors
> >
> > ➤ demobilization and re-mobilization costs
>
> *Indirect overhead*
>
> The proportionate share of office rent, office supplies, office utilities, office equipment, advertising, professional fees, management salaries, technical services, estimating, selling, accounting, bookkeeping and clerical expense, business licenses, taxes (except income taxes) and insurance. The "Eichleay formula" is used by some courts to figure indirect overhead. First, find total home office overhead expense for some period, such as for a month. Then compare billings for the delayed job with billings for all jobs the contractor has under way for the same period. Multiply that percentage by total home office overhead for the period. That's the "unabsorbed overhead" for the delayed job for the period. Convert that cost to a daily rate. Multiply the daily rate by the number of calendar days of delay. The result is likely to be a big number.
>
> But note that all claims for delay damages require proof of some loss to the contractor. Specifically, Eichleay-formula damages apply only if the contractor is standing by — ready to resume the job — and not fully occupied with other work.

Here are situations that have resulted in an award of delay damages to the contractor:

> ➤ an error or omission in the plans or specs that isn't (or can't be) fixed promptly

- a decision by the owner to change the scope of work
- a decision by the owner to suspend work before completion
- a failure of the owner to comply with the construction schedule
- a failure of the owner to yield control of the jobsite to the contractor
- a failure of the owner to make payments when due
- delay by a third party for the benefit of the owner or within the control of the owner

Avoid Delay Claims

Nearly all states either void or restrict enforcement of no-damage-for-delay provisions in contracts for non-residential construction. But that shouldn't give contractors a blank check if there's a delay. A CM contractor's best defense against delay claims is vigilance. When you sense that work is being slowed by the owner, try to get back on schedule. Advise the owner that there's risk of a delay claim. Then raise the issue of delay with the contractor. Ask for a written notice and a detailed explanation if the contractor intends to press a delay claim. Don't let a contractor wait until project closeout to make a claim for delay. Insist that grounds for any delay claim be put in writing during the delay. A contract clause such as the following is likely to be enforced in many states and for many types of work:

> *Within 10 calendar days after suspension, delay or interruption, contractor shall submit to owner a notice of claim which: (1) Explains fully all grounds for the claim, (2) Provides complete documentation supporting the claim, (3) Cites the day the delay began and the last day of the delay, if known, (4) Specifies the compensation requested, and (5) Documents each element of the requested compensation. Failure to give notice of either the inception or the termination of delay or failure to present a claim within the times prescribed, shall constitute a waiver of any claim for extension or additional compensation based on that cause.*

The best defense against delay claims is a good offense. Here's what I do to head off that kind of stuff. Every time you hold a meeting on-site, insist that delay days be the first topic of discussion. Record in the minutes of the meeting any mention of delay. Settle the issue of delay claims and change orders on the spot. Don't carry claims forward week after week. That breeds uncertainty and distrust. All claims have to be settled; the sooner, the better. Don't let future amnesia skew the odds against the owner getting a good settlement of the issue.

If you can't settle a claim in a weekly job meeting, take the contractor or superintendent out for a walk around the jobsite. Get the contractor or superintendent away from the audience common at jobsite meetings. After a little small-talk, get down to resolving issues. Offer to swap the claim for favorable treatment on the means and methods on some other issue. Do a little horse trading. If that doesn't work, dicker back and forth the cost of settlement. You're better prepared to do that than any two law firms.

To settle claims, you'll need the unwavering support of the owner. Committing the owner's money on the spot and without consulting the owner is a privilege you have to earn. If the owner doesn't have enough faith in your judgment to grant that privilege, ask for a dollar amount you can commit when settling claims. I've been granted as much as $15,000 as a settlement fund. You may need only a few hundred. Either way, make that money last. Return as much as you can to the owner at project closeout.

Delay Claims and Liquidated Damages

Expect hard bargaining over causes for delay and delay claims if the construction contract includes a liquidated damages clause. Liquidated damages identify a specific dollar amount to be charged for each calendar day of delay beyond the scheduled completion date. That makes sense when, for example, an owner will suffer a loss if a store doesn't open on time. We've seen liquidated damage clauses that range from a $1,000 a day for an office building to a $1,000,000 a day for a Las Vegas casino.

There's a good reason for specifying liquidated damages in the contract. The exact loss for late completion may be difficult or nearly impossible to calculate. A contract that sets a value on the expected loss (liquidated damages) makes those calculations unnecessary.

Courts in all states will enforce a liquidated damages provision in a contract if the damage amount bears some reasonable relation to the amount of the loss. Courts won't award liquidated damages if there's no actual loss or if the damage amount is considered a pure penalty. You'll recognize a liquidated damages clause in a contract from the words *time is of the essence*.

LEED Green Projects

LEED (Leadership in Energy & Environmental Design) standards promote efficient use of resources through better building design, construction and operation. A building can qualify for one of four LEED certification levels (certified, silver, gold or platinum) based on a point system, which considers site management, conservation of resources, and material selection.

The decision to apply for LEED compliance with the U.S. Green Building Certification Institute standards will usually require both changes in construction site management and changes in the materials ordered. LEED certification always requires that someone collect and report data that confirms LEED qualification.

Both residential and commercial projects can qualify for LEED certification. Most LEED points (referred to as "credits") are awarded for meeting design standards. But LEED credits are also earned during construction by (1) reducing waste and pollution, (2) recycling construction debris and (3) favoring certain types of materials. You'll have extra responsibilities any time the project is intended to qualify for LEED certification.

For other than residential buildings, an independent commissioning authority (CxA) leads the certification process. The CxA develops design specs intended to meet LEED requirements for certification. A different CxA may monitor actual construction. When construction is complete, the CxA will verify compliance by writing the commissioning report which qualifies the project for LEED certification.

Both single-family and multifamily homes, and both new construction and gut-rehab of existing dwellings, can qualify for LEED certification. A LEED for Homes Provider will review the plans before construction begins and arrange for inspections during construction by a Green Rater. When the job is complete, the Provider will submit a final LEED checklist to the U.S. Green Building Certification Institute.

As a CM contractor, you're an advisor to the owner. Unless you've spent quite a bit of time working on LEED projects, a word of caution may be in order. Owners love the thought of savings from LEED design. But there's extra risk and expense in every LEED project. Design, materials, inspections and certification will cost more. It takes a long time to offset $500,000 in added design and construction cost with $1,000 a month in reduced energy bills. More often than not, the owners will love the concept of LEED more than their budgets will. Initial enthusiasm over LEED design usually turns to disappointment when bids are opened. Do your LEED research and advise your client wisely. Many owners can afford LEED skylights but not super-efficient windows and not carpet made out of old tires.

"In the interest of safety, Coleson, may I make a suggestion?"

Jobsite Safety

A construction site can be a dangerous place. Every CM contractor has an incentive to keep the jobsite accident-free. Every CM contractor also has the responsibility to protect the owner from claims for injury on the construction site.

On larger jobs, the contractor will have a safety program, probably to satisfy requirements of the worker's compensation insurance carrier. It's part of the CM contractor's job to monitor that safety program. The plan should be a written document, should be available to supervisory personnel, and should be enforced.

Accidents and unsafe conditions should be documented and reported. The plan should include a list of required safety equipment by trade — hard hats, safety clothing, eye and hearing protection, masks, respirators, gloves, boots, guards and grounding for electrical equipment, fire extinguishers, etc.

In general, when an owner gives up control of the jobsite to a contractor, that contractor assumes responsibility for safety on-site. That's the way it should be. Responsibility for safety should rest with the person or company most able to control unsafe conditions. The contractor controls the jobsite. The design professional controls the plans and specifications. The owner controls the site before work begins and after work is completed. But there are gray areas where control may not be clear.

Any unsafe condition that exists on-site before work starts will be the responsibility of the owner, even after work begins. So it's a good idea to check for safety issues on-site before the contractor arrives. Once work has started, the owner won't be responsible for accidents due to conditions created by others. But there is an exception every CM contractor should understand. An owner (or a CM contractor) who exercises control over safety on-site may be taking responsibility for safety. Taking the initiative on safety issues can mean that you (and the owner) have taken some measure of responsibility for accidents on-site.

Still, it's poor practice to ignore an obvious safety issue simply to avoid legal liability. There are perfectly good ways to improve the safety climate on-site without taking responsibility for jobsite safety. Don't ignore a serious danger to health or safety. Instead, notify the contractor orally and follow up with a written confirmation. Explain that you feel the unsafe condition constitutes non-compliance with the contract, and that if the safety violation isn't corrected immediately, the owner intends to stop all (or part) of the work until the safety problem is corrected. If work is stopped, the contractor won't be entitled to an adjustment in the contract price and the contract time won't be extended.

Depending on what the contract says, you may not be able to enforce those terms. But it's unlikely that you'll ever have to. No contractor wants to lead a crusade for unsafe working conditions.

The issue is more difficult when a contractor claims that something on the plans or in the specs creates an unsafe condition. The best response is to request a proposed change order from the contractor.

1. Exactly what change would be required to resolve the issue?
2. What's the cost of making that change?

> **A CM Success Story**
>
> A few years back, I replaced a program manager group that had been fired by a school district. The project consisted of work on 34 separate sites, with a budget of $380,000,000. When I took over, six schools were under construction. The remainder weren't even being designed yet. Contractors on all six active sites were ready to sue the district over various delay claims and change orders. I soon found out why. Senior managers of the group had been promised a Christmas bonus if contractor claims came in below 2 percent of the contract price. To stay under 2 percent, the managers simply denied nearly all legitimate claims. That didn't sit too well with the contractors. Their attitude was, "We'll see you in court."
>
> I sat down with each general contractor over a cup of coffee and explained the situation. The district had no intention of cheating anyone out of a single dollar. If they had a legitimate claim and could back it up with paperwork, they would be paid. Most of their claims centered on lack of timely information and unforeseen conditions. It took me an entire year to wade through those claims.
>
> Next, I met with each of the district's architects. The previous management group had kept design personnel out of meetings with the owners. The architect's chief complaint was that the district was painfully slow in responding to inquiries about design. Progress in design work suffered. I told the architects that as far as I was concerned, they had a contract directly with the owners and should meet with them face to face. All I wanted from the design team was a weekly status report, a set of usable plans, a building permit, and an explanation of why they weren't done yet. We had contractors waiting.
>
> It must have worked. All 34 sites were built out, very nearly on schedule. And the district never got sued.

3. How long will it take to make the correction?
4. Why wasn't this issue raised before award of the contract?

When you have responses to these questions, it's time to get expert help from an engineer or an accident prevention specialist. Electing to proceed without a change order may mean that you accept liability for any accident.

Many accidents are the result of clutter and haphazard work. A clean, orderly site is a safe site. Insist on it.

Defective Work

Part of monitoring the job will be identifying work which doesn't meet contract requirements. No contractor wants to tear out work that's been completed. But if a tear-out is required, the best time to make that change is usually right away, and certainly before more of the same work is completed. Defective work identified early won't become the subject of a callback or warranty claim after project closeout.

Everyone can agree that defective work has to be replaced at the expense of the contractor. The dispute will be over what constitutes defective work. Obviously, anything that doesn't pass inspection or that fails a required test is defective. But most claims about defective work won't be that easy to resolve.

The contract may include guidance on correcting defects. Some construction contracts give the architect, engineer or owner absolute authority to decide what's defective and what's not. Courts will usually enforce a contract provision giving someone authority to identify work as defective so long as decisions are made in good faith. It's good practice to give the contractor both oral and written notice of defective work and then wait a reasonable time for a response. If the contractor feels work completed complies with the contract, that response will be a request for a change order. The response to that request for a change order may be another change order, this time from the owner, deducting the owner's cost of fixing the defect from the amount remaining due on the contract. At that point, the battle lines are drawn. The only choice may be to invoke the dispute resolution procedures provided in the contract.

Sometimes the defect is a minor cosmetic flaw in an otherwise-acceptable installation — such as a small chip or scratch in a granite countertop. Will the contractor be required to replace the entire countertop? If cosmetic repair is possible, suggest that the contractor accept a charge against the contract price equivalent to the cost of making cosmetic repairs.

Quality Assurance

The purpose of a QA program is to ensure that work complies with the contract, the building code and good professional standards. The QA program adopted by the contractor should be followed by both the contractor's personnel and subcontractors. The QA program will be more important under fixed price contracts where the contractor has an incentive to minimize costs. On larger jobs, the contractor may be required to retain an independent QA firm to conduct inspections, reviews and tests, to document non-conforming work, and then ensure that corrections are made. It's generally best if QA personnel work independently from installation personnel. Everyone works more effectively when they know someone in authority will check each step.

An effective quality assurance (QA) program can save money and time by reducing the risk of failed inspections and claims. Use the acronym PLANS to remember the essentials of a QA program:

Preparation (trained personnel, the right tools and materials)

Lists (installation checklist or instructions for each task)

Accountability (a record of who does the installation; who checks the work)

No adverse conditions (no excuse for defects)

Sign-off (by a supervisor when installation is complete)

Hazardous Materials

Hazardous materials fall into two categories: (1) those found on-site and (2) those brought to the site during construction. Generally, the owner will be responsible for hazardous materials *found* on-site. The contractor will be responsible for hazmat *brought to* the site. Dealing with hazmat brought to the site is probably the easier of the two. See the sidebar, on the following page, for a list of materials considered hazardous and that should be restricted on-site.

Manufacturers of hazardous materials publish material safety data sheets (MSDS) that explain how the material should be used, handled and stored. The contractor should provide the owner with a MSDS for any hazmat brought to the site. When in doubt about a hazardous material, refer to the MSDS.

Examples of hazardous materials that may be discovered at the site include asbestos, PCBs (in transformers), toxic mold, toxic chemicals, industrial waste, and radioactivity. The property owner will be responsible for hazmat found on-site unless it's brought to the site by a contractor or subcontractor. But the contractor will be responsible for any negligence in handling or removal of hazardous materials.

A contractor who finds what may be hazmat at the site should simply stop work in the area affected, keep construction personnel away, and notify the owner's representative. Unless specified in the contract, a contractor isn't required to do work that involves hazardous materials. That's work reserved for licensed specialists. Any contractor who runs into hazmat and continues work and spreads hazmat all over the site needs a lawyer, quick — and he better get out his checkbook.

Cleanup

The property owner's interest and the contractor's interest run parallel in most respects. Both want to see the project completed successfully and on schedule. The same isn't necessarily true when the issue is accumulated construction debris. Owners want to build good relations with neighbors and local authorities. A contractor's interest is likely to be much more limited. Sometimes contractors and subcontractors need a reminder about their responsibility for waste disposal.

Keeping the site clean and relatively free of debris is good construction practice. A clean worksite is also a safer worksite. A daily cleanup, or weekly, such as every Friday, can make site cleanup almost automatic, especially if debris boxes with plenty of capacity are near the points of installation. A project intended to qualify for LEED certification will have separate receptacles for waste products to be recycled. But nearly any project can benefit from separation of waste by category — such as corrugated paper, lumber scrap, concrete, asphalt, masonry, roofing, insulation, piping, wire and metals.

> **HazMat Commonly Brought to Construction Sites**
>
> *Asbestos*
>
> Materials with friable (airborne) asbestos are the most dangerous. Asbestos fibers released into the air can pose a significant health risk to anyone nearby. Non-friable (in mastic and glue) asbestos is less likely to be a health risk because the fibers are bonded into the base material. Some types of roofing, pipe and rigid siding include non-friable asbestos. But these materials can release asbestos fibers if cut or damaged. It's good practice to prohibit bringing any material with asbestos to the jobsite without prior approval. And approval is unlikely. Almost all products with asbestos are prohibited by either design specifications, product liability insurance, or state or federal laws.
>
> *Explosives*
>
> Many explosives are regulated by state law or local ordinance. For example, a permit is required to use blasting materials in residential areas. With the exception of powder-activated tools, it's best to prohibit bringing explosives to the site without a permit. Storage of explosives on the jobsite should be prohibited. Even if approved in advance, any loss or damage that results from use or handling of explosives by the contractor should be the responsibility of the contractor. Many materials release vapors that can explode when mixed with air. Don't allow large amounts of fuel, paint, solvent or toxins to accumulate on-site.
>
> *Lead*
>
> Except as required by contract. Lead-based paint is still made and has limited applications on some types of industrial projects. But don't use lead-based paint on residential or commercial projects.
>
> *Flammables*
>
> Flammable liquids with a flash point of 110 degrees F. or below should be stored in safety cans with the appropriate Underwriters Laboratories label. Paint thinner, gasoline, oil and roofing materials should be stored in a metal shelter at least 50 feet from any permanent structure. Storage drums for flammables should have a vented pump, not a spigot. Flammables such as gasoline, solvents and benzene can't be disposed of in a sewer or storm drain. Oil-soaked rags should be stored in sealed metal containers. Temporary heating equipment should have an Underwriters Laboratories label and combustion controls. Welding, flame-cutting and removal of paint with heating equipment requires a permit and should be approved by the owner in advance.

Most construction contracts give the owner the right to do a site cleanup and charge the contractor. Usually, a reminder of the owner's right to clean the site is enough to get cooperation from a contractor. Occasionally, an owner may have to take direct action, such as in response to complaints from neighbors or civic authorities.

Cleanup falls in two categories, progressive (as work is going on) and final (preparing the premises for occupancy). Final cleanup is usually done by a crew hired for just that purpose. Progressive cleanup is usually done by the same crew that did the installation. Subcontractors should be responsible for removing their own debris. Identifying the subcontractor responsible for any particular trash pile is seldom a problem. But it's the contractor's problem. It shouldn't be your problem.

The contract should set standards for both progressive cleanup and final cleanup. The responsibility of the CM is simply to see that those standards are met. Contracts available for free download at http://PaperContracting.com include clauses that cover both progress cleanup and final cleanup.

The Last Word

Several times in this chapter you've seen references to a potential for dispute. We're in favor of resolving claims on the spot, and as soon as possible, by compromise and horse trading. But some problems can't be resolved without assistance from others, such as courts or arbitrators. Every CM contractor has a responsibility to avoid disputes likely to inflate costs or delay completion.

As a general proposition, everyone has more to lose from a bitter dispute than they have to gain from winning an argument. As a construction professional, you have an advantage in resolving disputes. You're aware of conventions and procedures common in the industry. The property owner probably doesn't have the benefit of that experience. You know what's expected of property owners and what's expected of contractors. You're in a good position to resolve most disputes because you don't have a direct financial interest in the outcome. It should be easier for you to be objective.

Of course, the last resort in resolving any dispute is always the court system. But there's seldom a clear winner when a case has to be decided in court. The most common outcome is a settlement that pleases no one. Most of the money that changes hands goes to the attorneys.

Formal arbitration is usually a quicker, cheaper way to settle disputes. Smaller disputes (under $10,000) can usually be resolved by submission of documents alone. But even arbitration isn't cheap. It's much better if you can resolve disputes without resort to courts *or* arbitrators. Here's a procedure that works for many contractors:

When you've reached an impasse, suggest to both the contractor and the owner that disputed items be reserved for settlement when work is complete. The owner should pay any amount not in dispute and the contractor should complete any work not in dispute. When work is done, consolidate resolution of all disputed items into a single conference. You'll probably discover that some issues can be resolved with an exchange — mutual agreement to swap claims. Then get the opinion of a neutral third party or recognized expert on what remains. Most would call this informal *mediation*. The result isn't binding on anyone. But it gives both parties to the dispute an opportunity to state their case and understand the strengths and weaknesses of the opposing position. With that information, and with work complete on the project, most disputes lose their emotional component, making compromise the most attractive option.

Construction isn't like love or war. It's just business. And it's good business to find ways to compromise conflicting interests.

Chapter 8

Keeping the Owner Informed

THE LAST CHAPTER WAS mostly about collecting information, staying on top of day-to-day operations, both on the jobsite and off-site. This chapter and the next two chapters are about passing that information on to others. This chapter covers communicating with your boss, the owner. Chapter 9 covers processing payment requests, again with the owner. Chapter 10 covers communicating with the trade contractors and suppliers.

Most of what you pass on to the owner will be in the form of regular reports and summaries, some written, some by email, text message or phone, and sometimes just paper documents created by others. As emphasized in earlier chapters, the owner makes the decisions. As a CM contractor, you're "only" a consultant. You advise and recommend. You also have to provide the owner with enough information to make good decisions.

In the last chapter, we explained that monitoring day-to-day construction is, in many respects, the same whether you're a CM contractor, a prime contractor or project superintendent. Collecting information is about the same for any of those three roles. But what a CM contractor does with the information collected is different. Prime contractors and superintendents make decisions and give instructions — often immediately and on the spot. CM contractors make recommendations, sometimes at a later time and often not at the jobsite. That makes a CM contractor's role more difficult, at least in some respects. And it can delay job progress if not handled correctly and efficiently. We'll cover both the difficulty and the delay in this chapter.

> **No Ticket; No Laundry**
>
> Not too long ago, I was the CM contractor on a school job in California. The project was financed by both state and federal grants. What could be more secure than that? Surprise! Halfway through the job, California ran out of cash. Educational grant projects went on hold until Wall Street banks agreed to lend the state more money. That took eight months.
>
> Our job was shut down, fenced off and mothballed. Everyone on the job was sent home, including me. When the project started up again eight months later, the construction industry had collapsed nationwide. Half of our subcontractors had gone out of business. The general contractor had cut staff by 60 percent. About 90 percent of the construction personnel on my job were new faces. The contractor replaced his superintendent seven — count 'em — seven times before we completed the project.
>
> If you do nothing else carefully on a project, make sure the money's there in the beginning. Even exercising extreme care, the mysterious world of high finance can get in the way of progress.

Heading Off Problems

Talk to an experienced CM contractor for very long and you'll probably hear a story or two about jobs that went bad, often because of friction between the owner and the CM contractor. The owner and the CM contractor are on the same team. Their positions are mutually reinforcing. The CM contractor has to rely on the owner and the owner needs confidence in the CM contractor. When relations between the two break down, it's almost inevitable that job progress will suffer. Contractors and subs will be among the first to detect signs of friction. Clearly, it's important that you maintain good relations with the owner. A little planning should head off the most serious problems. If you can manage the following six issues successfully, you're likely to avoid the most common disputes that develop between CM contractors and owners.

1. **Information:** Keep it flowing through appropriate channels. Get a commitment from the owner to respond promptly and fully to your requests for information. You're not going to bother the owner with trivia. Your request for a decision will almost always come with a recommendation. Deciding shouldn't be too hard or take too long. The default decision is always a simple "OK." So when you need a decision, your request should be a top priority for the owner. Any time the owner isn't available, someone should be authorized to make decisions on behalf of the owner. Once a decision is made, you're entitled to rely on that decision. This usually means important decisions should be in writing. If the owner doesn't want to respond to questions in writing, no problem. You'll send a summary of the decision by email or text message within 24 hours.

"Why not use the $1,600 a day we're collecting for delay to send my crew on a Caribbean cruise?"

2. **The owner's authority:** Be sure there aren't any legal reasons why the owner can't make final decisions. If there are multiple owners (such as husband and wife, partners, or a board of directors), is the owner you're dealing with authorized to speak for all co-owners? For example, if you're working for a board with six or more members, two should be authorized to make decisions for the entire board. Those two board members can debate their decisions with the full board until the cows come home. It won't affect progress of the job. If you have any doubt about the owner's authority, ask to speak with the spouse, partners, or members of the board of directors. If you're doing a project for a tenant, the lease agreement probably sets limits on improving the property. Ask to see a copy of that lease, or to speak to the landlord.

3. **Money:** Talk about it. Get the owner into a candid conversation about where the money is coming from. If the project is financed out of a bank account or stock portfolio, ask to see a statement. If there's a loan from a commercial lender, ask to see that commitment. Be especially wary if the source of construction funds hasn't been determined yet. Your commitment to contractors and suppliers is no better than the source of the owner's funds. Satisfy yourself that the owner has the resources available to see the project through to completion. Funding isn't everything, but it's the only thing that keeps construction moving.

4. **Delay:** Discuss the costs associated with delay by an owner. The Martin Brothers vs. VMI case in the last chapter is a good example — $1,600 charged per day of delay. Get a commitment from the owner not to hinder or delay the work except as permitted by contract. If any architectural review committee or owner's association has the right to make demands before or during construction, get a commitment from the owner of "best efforts" to meet requirements of that committee or association.

5. **Separate contractors:** If the owner will have separate contractors working independently on the site, get a pledge from the owner that your work can proceed without interference. The owner's agreement with separate contractors should prohibit any activity which would delay your work.

6. **Materials:** It's common for owners to order and pay in advance for any materials or equipment that require a long lead time. When the owner arranges for materials, those materials should be available when and where required in the normal course of construction. Be sure materials ordered by the owner will be suitable for the intended purpose and won't delay completion. Try to avoid surprises on topics such as shipping charges, damage claims, insurance, warranty, delivery instructions and storage fees.

If the owner isn't savvy about construction scheduling, get the owner to select, purchase and store any owner-supplied materials well before installation is required. Don't discover, for example, two days before installation is planned, that carpet hasn't yet been ordered. Every CM contractor's primary responsibility is to keep the project under control and moving at all times... Period!

Certainly, there are other potential issues which can disrupt harmonious relations between a CM contractor and an owner. But resolving these six potential issues early in the project cycle should help promote smooth operation through completion.

Information Channels

Construction can be chaotic — an endless series of questions, urgent requests, corrections, and minor emergencies. Some issues require a decision before work resumes. Fortunately, modern technology makes it easy to pass information and requests almost immediately. Unfortunately, easy communication can be both an advantage and a disadvantage. The advantage is obvious. The disadvantage: It's not easy to keep information under control.

Here's an example: The project owner makes all the final decisions. We know that.

Separate Contractors Working to Your Advantage

I did a condominium remodel in Hawaii recently. The owner hired a local general contractor to run the job. I did the design work and represented the owner. Under the contract, we had the right to hire a separate contractor to complete any part of the job.

The tile contractor's bid came in sky high, at least $10,000 more than the job was worth. I couldn't get him off his price. Maybe there was a kickback somewhere. Or maybe tile subs in Hawaii have an agreement on splitting work. I don't know. So I took another approach. After consulting with the owner, I told the contractor we were invoking our right to use a separate contractor. The owner would buy and install the ceramic tile. That left our general contractor a tad grumpy. But there wasn't anything he could do. We had the right under the contract. But he gave me a window of just one week to finish the tile work.

I called a mainland contractor I know and asked him to recommend a crew of tile setters. I offered to cover airfare and living expenses in Hawaii in return for a week of work. We had no trouble recruiting qualified tile setters. I flew the crew to Hawaii, rented a car and put them up in a condo adjacent to our site. Less than a week later, the job was done. The tile crew had a couple of days of play time on the sand, and I saved the owner nearly $8,000 on the deal.

Not every separate contract will work out so well. If the owner has the right to supply labor and materials, get a very clear understanding on when, where and how that work will be done. Get that agreement before work starts. Mid-contract is much too late to sort out these details.

So does everyone else on the job. Should the electrician or the plumber call the owner every time there's a question? Of course not. The owner may be eager to help, but isn't likely to have the background required to make good decisions on construction details. Any instruction an owner gives an electrician is likely to conflict with instructions from the electrical subcontractor, the prime contractor or the CM contractor. Obviously, the best flow of information is through lines of authority. We call those *channels*.

Early in the project, every CM contractor needs to set standards and procedures for the flow of information. No one will feel compelled to go outside authorized channels when requests, questions and documents are handled promptly and efficiently in proper channels. The owner should understand both your reporting system and that there are valid reasons to stay within that system. Trade contractors need to understand that you're the point person for their questions and requests. They have a contract signed by the owner. But you've been appointed by the owner to respond to requests from contractors. You're ready to do exactly that. There's no need to contact the owner directly.

Staying in Channels vs. Developing Consensus

Owners like to be helpful. Most won't deliberately undercut the authority of their CM contractor. But an owner on-site during work hours is going to be asked questions and is likely to respond. That response carries the weight of law, even if it's just a casual remark made without a lot of thought. When something like that happens, it's time to have lunch with the owner. You need to get the pecking order back on track.

An owner or designer who's lost faith in your control of the job is likely to circumvent lines of authority. Here's an example. I was the program manager on a large educational facility in central California. The project was still in the design phase, and the designers were making good progress. Several instructors had been appointed to advise the design team, but their ideas were being ignored, and they weren't happy about it. I explained to the design team that they needed to work harder to satisfy the instructors.

That didn't set well with the designers. They went straight to the superintendent without my knowledge or consent. Once that came to my attention, I went to see the superintendent and explained the breakdown. The architect's proposed design didn't address issues the instructors considered important. I felt the designers could do better. But we needed his (the superintendent's) support. He agreed and recommended that I sketch a design that resolved the instructors' concerns. I did exactly that. With several instructors present, we presented design alternatives to the full board. The board selected my design over the original architect's design, and that's what ended up getting built.

CM contractors occupy a very privileged position. You don't design anything and you don't build anything. Any architect or general contractor would love to have a job like that, with plenty of cash flow and no risk. It's like running a diamond mine without the trouble and expense of digging.

But with those benefits come a burden: Every CM contractor is expendable. Any architect or contractor on the job is in a position to step into your shoes. Inept CM contractors get replaced. And the easiest way to appear inept is to be somewhere else when needed on-site. Make it your practice to be highly visible every day, either in the owner's office or on the jobsite. Out of sight is out of mind — and out of a job. Leave a vacuum and the architect or general contractor will fill that void. If you're trying to run 10 projects for 10 different owners, count on losing at least two or three of those jobs. It's far better to have two good-sized projects you can control than five you can't.

> *"If you're trying to run 10 projects for 10 different owners, count on losing at least two or three of those jobs."*

The remainder of this chapter fleshes out an information processing system that should provide good service on nearly any construction project, particularly a project run by a CM contractor. One word of caution: What's required on any particular project depends on job size and complexity and the personnel involved. For the smallest residential job, you'll need only a small fraction of what's recommended here. For a larger project, you'll probably want to supplement the system outlined with more reports, deadlines and specific guidance. For the largest jobs, expect the owner to provide a written construction management plan which identifies how project information should be routed.

What's in Those Channels

To this point, we haven't defined *information*. You'll understand where we're headed when we get more specific. All of the following constitute *information* for which there should be an appropriate channel:

1. schedule updates, including status reports on progress during the reporting period
2. permits, submittals, samples, shop drawings and testing reports
3. correspondence, conversations and meeting minutes
4. potential delays and disputes
5. budget updates, payment requests, certified payrolls, cost summaries and cash requirement forecasts
6. requests for change orders and clarifications
7. close out information, training, operation manuals, warranties, as-built drawings and specs

Who creates this information? Who gets the information and how? What response is appropriate and when? Where should documents or files be stored or archived when processing is complete? We'll cover items 1, 2, 3 and 4 in this chapter. Processing payment requests is the primary topic in the next chapter. Chapter 11 covers change orders, and Chapter 13 covers project close-out.

Status Reports

Even if the owner is a regular visitor on-site, it's good practice to provide regular status reports. A status report confirms that you're on the job and making progress. The best time to submit this report is usually when you present an invoice. A status report can take almost any form. On a small job, it's probably enough to write a paragraph or two summarizing what's happened on-site since the last report. On a larger job, you'll need more detail: lists of work started and work completed during the period, the results of tests, inspections and changes in the plans or schedule. If the owner hasn't been on the jobsite since the last report, include pictures that document progress. Legitimate topics to cover in status reports include:

> **Schedule:** Changes since the last status report. Include an updated network diagram (CPM) or bar chart if you have one and a new estimated completion date if there is one.

> **Costs:** Percentage of the contract price incurred to date. Provide a new estimate of the completion cost if any change orders have been approved since the last progress report.

> **Problems:** It's better to be candid about challenges, both anticipated and overcome. No one likes to deliver bad news. But if you have some, share it sooner rather than later.

> **Solutions:** Identify steps being taken to ensure completion of the project on scheduled and within budget.

Email is usually the best way to deliver a status report. A copy will be archived automatically in both your outbox and in the inbox of the recipient. But don't rely on email exclusively. Hand deliver a weekly progress update personally. That's the best way to stay in front of your owner.

Submittals, Samples, Shop Drawings

Many "approval" items require consent by the designer, owner or owner's representative before the material is purchased, fabricated or installed. Some submittals are required because the designer elected to leave design details to others later in the construction process. Shop drawings fall into this category. Other submittals simply confirm that what the contractor plans to install will meet contract requirements. For example, the contract may require that product performance data or a product warranty be approved before installation.

"Can't you just make a general petition for timely approval of all shop drawings?"

Detailed product specifications are important. But the model number of a sink won't mean much to an owner. Pictures are better. Most submittals contain lots of cut sheets and technical data. But a nice little picture of the sink you're buying for the project will mean much more to Mr. and Mrs. Owner.

To identify when submittals, samples or shop drawings are required, check the plan details and job specifications. On a larger job with many approval items, the design team should create a comprehensive list of approval items. The contractor will need a similar list with a deadline noted for submittal of each item and the days allowed for approval. The CM contractor should verify that the contractor's approval list is complete — nothing omitted.

Keep a submittals log that numbers all of the submittals; who they came from, when they were due, when they arrived, when they went out for review, when they came back, when they got reworked, when they went back out for re-review, when they came back approved from the designers and when they went back out to the general contractor. This log will be worth its weight in gold if a dispute about submittals goes to a settlement conference.

On a larger job, submittal procedures will be laid out in detail in the contract — number of copies, submittal format, routing, etc. But even on small jobs, it's good practice to have written rules for handling approval items.

For example:

> ➢ **Letters of transmittal** that accompany approval items should identify the date of submission, what's being submitted for approval, when approval is required to stay on schedule and the revision number, if any.

- **Copies** of submittals should include enough for distribution to the designer, the owner, the CM contractor, the prime contractor and the subcontractor or vendor who prepared the submittal.

- **Shop drawings** will be prepared by manufacturers or fabricators to show details about some material ordered for the job. For example, if custom-fabricated casework is required, a cabinet shop may have to submit plans for approval before the cabinets are built. The drawings should show installation details and how work will be coordinated with other trades. If the shop drawing includes structural items, engineering calculations and the engineer's seal should appear on the calculations. Shop drawings for items such as aluminum store fronts and fire sprinkler systems will require separate permits.

- **Samples** can include almost anything required for the project: window and door sections, hardware, flooring, fixtures, wall treatments, etc. Many samples are simply color chips or examples of texture or workmanship.

- **Samples** should be boxed or wrapped and labeled with details about the item: brand and model number, name of the manufacturer, who submitted the sample, on what date, and who received the sample.

- **Contractor review:** Submittals prepared by subcontractors and vendors should be reviewed and certified as accurate by the prime contractor before being submitted for approval. The contractor should sign, stamp or initial every submittal prepared by a subcontractor or vendor.

- **Non-complying submittals:** On most jobs, it's OK to ask for approval of something that doesn't comply with the plans or specs. But the letter of transmittal should include a request for variation and any required change in the contract price or completion date. A non-complying submittal is called a *substitution submittal*. Do your research before trying to substitute materials. On large jobs, substitutions will be disallowed if you don't follow the procedures set out in the contract or the specifications.

- **Voluntary submittals** should be welcome. A contractor who isn't sure a color or texture or product spec meets the owner's needs should be able to get an approval before installation begins. On small jobs, no formal approval process will be required for voluntary submittals. But it's common courtesy to confirm that the voluntary submittal presents no problems.

- **Partial submittals** should be welcome. For example, if tile samples are required for both floors and counters, it may make sense to have samples approved separately.

- **Deadlines:** Submissions should be delivered early enough to give the owner or owner's representative enough time to issue an approval without delaying the project.

> **Rejection:** Any submittal that's rejected should include a clear statement of the grounds for rejection. Leave no doubt about what's needed for approval. Rejections should be marked up in red ink to indicate the changes needed.

> **Resubmissions** should identify what's been changed and include a copy of the statement of rejection issued with the prior submission. Include one copy of the original submittal in case the designers can't find their original — they're in the design business, not the filing business.

> **Certificates, inspections and tests:** When a particular task is completed, the specs may require testing or certification to demonstrate compliance with standards set for installation. This form of submittal should meet all requirements of other submittals: letter of transmittal, copies, contractor review, etc. On larger public jobs, the specs may require in-shop fabrication inspections and testing. A state-certified inspector may have to visit the fabrication facility to do the inspections.

Significance of Approval

A design professional working for the owner may not want to take responsibility for design work done by others, such as a shop drawing. Architects prefer to limit their responsibility for submittals to checking compliance with the *design concept*. That standard offers little comfort to an owner or contractor.

The architect's review of submittals can be either:

> *for design compliance only* — the architect offers no opinion on whether the submittal complies with the plans and specs

> *for contract compliance* — the architect confirms that the submittal complies with the plans and specs. The contractor will still be responsible for proper fit, installation and performance

The construction contract and the owner's agreement with the architect should identify which of these standards will be used on the job. Regardless of the standard selected, all can agree that neither the contractor nor the building owner should be required to practice architecture without a license. If there's a licensed design professional on the scene, that individual has an obligation to observe the standard of care appropriate for that profession when reviewing submittals.

The construction contract may specify the time allowed for approving or rejecting submittals, samples and shop drawings. If the contract doesn't cover the subject, five days will usually be a reasonable time. In some cases, approval may be conditional — approved if certain conditions are observed. That's usually a better choice than outright rejection.

On a larger job, the contractor should retain each approved original. At substantial completion, the package of approval documents should be delivered to the CM contractor along with an invoice for final payment.

Retaining Construction Records

What records should a CM contractor keep and what records can be discarded? No two CM contractors will have the same answer to that question. But most would agree that keeping every shred of paperwork isn't necessary.

Everybody on the project has files. The designer, engineers, contractors and subcontractors all have files. The owner may even keep some files. In the pre-construction meeting, hand out a memo specifying that everyone involved keep their own files current. Emphasize that those files should be *available to the owner and the owner's representatives on demand*. That shifts much of the burden of record retention to others. The documents you need most will usually be readily available in your inbox or on your desk. (You may need to clean your desk to find them.) If you can't find what you're looking for, request a copy from the designer or the contractor. Like magic, the required document is likely to show up in your email or fax machine. But keep everything with the potential to turn legal down the road. Start your legal file on Day 1. Watch that file like you'd watch a bottle of fine Scotch.

Anyone who's been involved in a construction dispute will agree that good records win both arguments and court cases. Contractors keep payroll records. Owners keep track of payments. CM contractors should keep records on:

- correspondence, written or emailed
- notes on conversations, directions and decisions
- digests of meetings and jobsite meeting minutes
- field notes, as-builts and dimensions
- inspections, tests, boring logs and reports
- requests for interpretation

A correspondence file is probably the easiest set of records to keep. Simply file paper copies by project name in chronological order. When the file has grown to more than fits comfortably in a manila folder, start several files, perhaps in alpha order by the name of the sender or recipient. Email correspondence is archived automatically in most email programs and can be sorted and searched in seconds. But nearly every job will also require a correspondence file for paper documents. Just to be safe, consider printing important email messages. The wonderful world of digital storage is far from perfect. In court, 10 pounds of paper will beat a corrupted microchip any day.

Most correspondence in your file will be from contractors, subs, vendors, inspectors, manufacturers or the owner. But copies of correspondence you send should also be kept in the paper file. If you use Acco fasteners in the folder, keep correspondence you send on the left leaf and correspondence you receive on the right leaf. Any time you send a paper memo, reminder, confirmation or receipt to someone, place a copy in the correspondence file.

When the file begins to bulk up, subdivide correspondence into categories such as design, permitting, and construction. If the construction file gets too thick, subdivide that further into concrete, framing, roofing and the like. Keep separate files for budgeting, scheduling, payments and change orders.

Keep notes on important conversations and meetings. We all have a convenient memory — remembering best what we want to believe. When you see something that has to be changed or corrected, document what you've seen or done and add that memo to your correspondence file — even if you don't send the memo to anyone. There's a good legal reason to follow this practice. Courts and juries give extra credence to notes created *at the time something happens*. The doctrine is called *contemporaneous business records*. As a witness, you'll be allowed to refer to notes you made at the time in question. Notes in your own handwriting are best.

Distribute notes or minutes kept at meetings to each person in attendance. Offer each participant the opportunity to review and correct the record of what that person said at the meeting. If there's no correction within three days, the minutes stand as written. When corrections have been entered, the notes or minutes should be distributed to all concerned and then filed for future reference.

Facts Win Cases

The rotunda at the Neiman Marcus store in San Francisco is a local icon. A hamburger, a glass of wine and the tip will run you about $35 at the bar under the rotunda. But that's a bargain. Windows that line the rotunda offer a view of Union Square well worth the price. Then look up to see a gorgeous stained-glass skylight spanning the entire space.

In 1984, I was the eighth project architect hired to restore this rotunda after years of neglect. When I got involved, construction was already under way. My job was to oversee design. If you haven't been to Union Square, I should point out that the Neiman Marcus rotunda is about same size as most state capitol buildings.

This job had issues, and I really mean all sorts of issues. The first issue was that reconstruction of the rotunda had to be done with the original broken plaster pieces that had been taken apart with crowbars and axes, trucked across the bay to Oakland on an unpadded flatbed truck and then abandoned in an unprotected public storage facility. There the pieces remained for 11 years while the zoning and plans for the store crawled through the public approval process. Over the 11 years, the public continually helped themselves to the classic ornate antique plaster pieces. That's when I showed up on the scene.

I did an inventory of the plaster parts and discovered that only half the rotunda was there. The other half of the rotunda had been distributed to collectors in Oregon, Ohio, Arizona or New Jersey. I did the shop drawings myself, set up a plaster fabrication shop and had crews start molding the missing plaster pieces. In the meantime, the general contractor was erecting the rest of the building all around me.

(Continued)

Facts Win Cases, *continued*

I had only one problem with the general contractor's company. They hated architects. So I dropped by their office at the same time every day to erode their resistance. It must have worked. Gradually, I took over anything that had to do with the rotunda. The only other problem: The original rotunda consultant took his money and split, leaving me to complete the project on a comparative shoestring.

The project was awash in design, construction and political intrigue. Any job like this attracts prima donnas — and their lawyers. I hated going to work. The entire site was thick with resentment, complaints and animosity. I couldn't possibly keep notes on all that. Then I had a brainstorm.

I went to Radio Shack and bought a small hand-held tape recorder and a bag of blank tapes. Every day, hour by hour, minute by minute, I recorded everything that happened on that job. I recorded descriptions of the work, the people, the meetings, the attitudes, the refusals to cooperate, every verbal dispute, theft of our new plaster castings — everything. I made what had to be the most complete audio record ever prepared for a construction project — in my own words.

We finished the job. It's beautiful. Go have a look the next time you get to San Francisco. But, as I expected, blame ran hip-deep at project closeout. Owners of my design firm asked me if I had any files or notes on the project, as they needed to cover their legal exposure. I was ready. For months, my secretary had transcribed those audio recordings. I went to my file drawer and pulled out a 4-inch-thick three-ring binder — over 200 pages of day-by-day, moment-by-moment, blow-by-blow accounts of the design and construction process as it actually happened. Just about everyone involved had egg on his face. That was the last we ever heard from opposing counsel. As a general rule of thumb, the person who is most organized and has the most complete information wins in court!

Today, making a verbatim video record would be both easy and cheap. When you anticipate trouble, I recommend keeping the best records possible. Good records will keep you off the courthouse steps.

Your Job Log

As a CM contractor, you're on the owner's team. Almost certainly, you'll be the most experienced construction professional on that team. If there's a dispute about something that happened during construction, your opinion and recollection will be very persuasive. In many cases, the owner will rely on you alone to support his position. Chapter 12 has more on what you can do to protect the owner's interest.

On any job, the CM contractor will have many opportunities to protect the owner from disputes. Facts win disputes and court cases. No one is in a better position to collect facts about the job than the CM contractor. The more complex the job, the more opportunities you'll have. Regardless of the size of the project, a written job log is the best way to collect and organize field notes. Think of your job log as the equivalent of ship's log. No ship would sail without one.

The best job log has bound pages numbered by the printer. That adds credibility: no pages have been either inserted or removed. Start a new page for each day. Record the date and the jobsite address. You can write anything you want about anything you want. But remember, anything you write can end up as evidence in court. So be cautious about what you commit to paper.

Note in your job log anything that could influence job cost or progress:

- weather (temperature and precipitation)
- crews on-site and the work done
- any work done on a cost-plus (time and materials) basis
- quality control issues
- tests and inspections
- deliveries received
- commitments made
- milestones completed
- schedule changes requested and granted
- change orders requested, granted or paid
- proposed and accepted design changes
- meetings and conferences
- decisions made or deferred
- requests for interpretation
- significant conversations or messages
- heavy equipment on-site — in use or idle
- visitors to the site by name, company and time
- who prepared the report
- who's been through security screening (LIFESCAN) background checks
- who has badges and grand master keys

Be especially careful to note issues that could turn into a dispute later — partial deliveries, conflicts between trades, incomplete or defective work, unsafe practices and accidents that result in personal injury. When a delivery arrives, open and inspect the contents *before the trucking company leaves*. To avoid problems later, note any discrepancies in your log book and make the insurance claim.

If you can't locate a book with numbered blank pages, almost any bound tablet, such as a laboratory manual, will work nearly as well. It's better if pages are bound into book form rather than loose leaf. A book keeps all job log pages permanently bound together in sequential order.

Contacts for Everyone

You can perform a real service by creating and distributing a list of contact numbers for everyone involved in the project. This project directory is like a membership list for a club or social organization.

Contact numbers should include:

- name and address
- role in the project
- phone numbers
- fax number
- email address
- alternate email address
- delivery address

Be sure to identify alternates; the person who should be informed or can make decisions if the primary contact isn't available.

Be sure everyone involved in the project gets a copy of your directory: contractors, subcontractors, key vendors, the owner and owner's representatives, design personnel, even the building department. Your name and numbers should be at the top of the list. Be sure emergency numbers are included. This project directory will become one the most valuable documents on-site. But the best part of this directory may be the cover page. That's where you set out a communication policy for the project.

Your communication policy should encourage questions. But, except for true emergencies, questions, invoices, documents, receipts, submittals, approvals, invoices — all information — should follow established lines of authority — from vendors and subcontractor to the prime contractor to the CM contractor and, if necessary, to the owner. Then the process reverses, passing payments, approvals and decisions from the owner to the CM contractor to the prime contractor and, if necessary, to subs and suppliers.

Your project directory will need regular revision. Keep the names and numbers current. Distribute directory updates when changes are required. Every update should include your cover sheet with the communication policy in force on the project. Some on the distribution list will need regular reminders to keep communication within lines of authority. Regular revision of the project directory can provide those reminders.

The best way to distribute your project directory will usually be as an email attachment.

Record Documents

Construction contracts for larger jobs usually require that the prime contractor provide the owner with a set of *record documents* at project completion. These record documents may include:

- as-built drawings and specifications
- submittals and test reports
- contract addenda and changes
- clarifications
- field changes and corrections
- layout of underground utilities
- coordination drawings
- your job log
- progress photos
- warranties and master keying information

Mechanical and electrical subcontractors usually prepare their own drawings for ductwork, supply, drain, waste and vent piping, fire protection piping and electrical raceway. These will be useful any time the building is changed or repaired. Make mechanical and electrical drawings part of the package of record documents delivered to the owner.

Some of these documents may be on paper. Others will be in digital form on disk or memory chip. Regardless, all can be converted to digital form for easy archiving.

No matter what the contract says or doesn't say about delivery of project documents, it's appropriate to transfer all project documents to the owner at project closeout. Both the prime contractor and the CM contractor are likely to feel a sense of relief the day their job file can be cleaned out and turned over to the owner. Some of these documents will have value during the entire life of the building. Others have little or no value after completion. Let the owner decide what's important and what can be recycled. What's clear is that neither the prime contractor nor the CM contractor needs to preserve project documents once work is done, final payment has been made and all potential disputes have been resolved. After that, the only remaining responsibility will be the general contractor's warranty period.

When you transfer record documents to the owner, be sure there's a written memo documenting the transfer. Get an acknowledgment of receipt — or at least send a memo confirming delivery of all project documentation.

Note that record documents don't include pay records or the original plans, specs and contract. You have good reason to keep these under your control after work is completed.

Some contracts require a survey of the site when all work is complete. Typically, a licensed land surveyor will be called in to certify that the project is entirely on a single lot, within any building restriction lines, and doesn't overhang or encroach on any easement or right of way. If a survey is required on completion, that survey should be among the record documents delivered to the owner.

The Money Comes Next

This chapter has explained a CM contractor's responsibility for keeping the boss informed – how information passes up and down the lines of communication. In the next chapter, we follow the money: payment requests routed from contractors and vendors through the CM contractor to the owner or lender, and deciding who gets paid, when, and how much.

Chapter 9

Evaluating Payment Requests

PAYMENT FOR CONSTRUCTION WORK is a big subject. There's a lot to understand, including what the law requires. Many states have prompt payment statutes which impose penalties on owners and contractors who don't pay on time. Lien laws on private construction projects help contractors and subcontractors enforce their right to collect. CM contractors are right in the middle of nearly every payment issue, acting as a gatekeeper, recommending that some invoices be paid and others be rejected. This chapter will explain what CM contractors need to know to avoid the most common errors.

A CM contractor's role in the payment process could be compared to the roles of both an umpire in a ball game (enforcing the rules) and an accountant (counting the dollars). In this chapter, we'll cover both roles. The umpiring comes first.

The rules are set by contract and state law. When is an invoice due? What has to be on that invoice? Who approves the invoice? How many days can pass between delivery of an invoice and payment of the amount due? What constitutes a legitimate reason for rejecting an invoice? What's the penalty for late payment? The contract and state law will answer these and dozens of other questions about payment.

Counting the dollars requires a recordkeeping system. Figuring the amount due can be easy on a small job with almost no change orders and payment based on job phases. But stray very far from that set of circumstances and the accounting gets more complex.

- On a larger job, there may be hundreds of change orders. On a billion-dollar casino project I managed, I had 600 change orders on the owner's side alone.
- Payments may be based on a schedule of values, with dozens of line items in each invoice.
- The amount due at any pay period can depend on the cubic yards of soil excavated or tons of steel erected.
- On a cost-plus (time and material) job with a guaranteed maximum price, the amount due may be a matter of opinion, not subject to precise calculation.

When the accounting gets complex, I hire a CPA to control the billings and bookkeeping. That makes it much easier to keep a firm grip on the economic paper trail. If you don't have the luxury of professional accounting help, an Excel spreadsheet will do most of what needs to be done.

The Request for Payment

Payment processing begins each time a CM contractor receives an invoice from a contractor or supplier. Occasionally a subcontractor or vendor may try to present an invoice directly to the owner, probably assuming that's the fastest way to get paid. Presenting an invoice directly to the owner simply adds an additional step to the process. The owner doesn't pay bills without your recommendation. You need to see each invoice, enter the numbers in your accounting system, review the invoice for accuracy, and make a recommendation to the owner. Insist up front, and in writing, that invoices be presented directly to you as the first step in payment processing. That's easier if the contract between the owner and contractors makes it clear: invoices require your approval for payment.

On public projects, invoice processing may be a little different. State law may require that the state's inspectors and designers approve each invoice before it hits your desk.

Evaluation of payment requests is one of the most valuable services any CM contractor performs for a property owner. It's the best protection an owner has against charges for non-conforming work and overcharges. It's also important if your fee as CM contractor is based on a percentage of the job cost. If you don't see all the invoices, you have no way of calculating the job cost. To protect yourself, include in your contract with the owner that you have the right to audit the owner's books.

Nearly all payments will be one of the following:

> ➤ an initial payment, sometimes called a *retainer*
> ➤ a progress payment, usually billed monthly or by completion of a specific project phase
> ➤ a final payment, including any remaining change orders, minus retainage
> ➤ a release of retainage, which may be paid in two parts, one at mid-project and one at completion
> ➤ a payment for a change order

Initial payment (the retainer). This is like a downpayment. It seals the deal, a sure indication that the owner is serious about proceeding. Initial payments are made before materials and equipment arrive on the jobsite and help the contractor carry the cost of mobilization. In some states and for some types of work, the amount of the initial payment is limited by state law. For example, the initial payment for home improvement work in California is limited to 10 percent, or $1,000, whichever is less. In Massachusetts, the initial payments for home improvement work can't exceed one-third of the contract price plus the actual cost of any special order materials or equipment.

Progress payments (monthly). If the job is expected to last more than a month or two, there will be progress payments. These payments help trade contractors carry the expense of construction while work proceeds. The contract will identify when payment is due. There are two common payment schedules. Progress payments can be due on:

> ➤ *completion of each job phase* — The sidebar outlines a common payment schedule.
> ➤ *a calendar date* — Trade contractors submit invoices for payment on a specific day or days of the month. That usually happens on about the 25th of the month, leaving five or six days for review and processing of the invoice before payment is due.

Final payment. This is the last of the progress payments and is usually due at substantial completion; the date the work is ready for occupancy. Any work yet to complete, such as cosmetic touch-up, will be on a punch list of items to be completed at a later date. The last payment on a job is due at final completion. That's when the punch list is worked off or when retainage is released.

You'll usually withhold from final payment an amount equal to 125 percent of the cost of completing the punch list. At this stage of the work, the contractor will probably be deep into some other project. Holding that 125 percent keeps your job a high priority in the mind of the contractor — providing an incentive to work off punch-list items. That 125 percent is probably a large chunk of the contractor's profit.

> **Progress Payment by Job Phase**
>
> A payment schedule like the one outlined below simplifies approval of payments. A schedule like this leaves little doubt about when payment is due. Most milestones in the schedule correspond with completion of an inspection. For example, when the foundation passes inspection, 10 percent is due. The amount due is a percentage of the contract price, less any initial payment and retainage.
>
> - 10 percent upon breaking ground, or when rough grading is complete
> - 10 percent when the underground utilities and foundation are complete
> - 15 percent when rough framing and plywood sheathing are complete
> - 10 percent when rough plumbing, electrical and HVAC are complete
> - 10 percent when doors and windows are installed
> - 10 percent when exterior walls and roof, insulation and finishes are installed
> - 10 percent when cabinets and countertops are installed
> - 10 percent when final plumbing, mechanical and electrical work pass inspection
> - 10 percent when exterior and interior finishes are complete
> - 5 percent when the job passes final inspection and receives an occupancy permit
>
> The flaw in this schedule is obvious. The percentages listed may or may not correspond with the actual cost of construction. Some owners would insist that the schedule is *front loaded* — the contractor collects faster than work is completed, and so gets ahead of the owner. For example, the contract could make 10 percent due on the first day of work. The contractor hasn't done anything more than show up, and he's already pocketed 10 percent! Any payment schedule that exceeds expenses removes part of the incentive to complete work promptly. But a payment schedule that trails expenses can leave a contractor strapped for cash needed to complete the project.

Release of retainage. Most larger construction contracts permit the owner to withhold some percentage (usually 5 or 10 percent) of each progress payment. Retainage will be released when punch-list items have been resolved and when there's no legal reason to withhold any portion of the contract price.

Retainage may be split in two parts, the first part being paid at about 60 to 75 percent completion. I don't recommend releasing half of retainage when the project is halfway done. That would sacrifice part of the owner's leverage. On large projects, retainage is usually kept in a trust account at a local bank. Interest on that account will be paid to the contractor on certification of project completion.

Change orders. Payment for smaller change orders should be due in full at the next regular billing after the change is complete. A change order that requires several months to complete may be billed out over several months. For example, I had a project where fire sprinklers were added by a change order in mid-stream. The same owner decided to move an entire staff department from one building under construction to another building under construction — requiring a major reconfiguration of space. Changes like that can span billings for many months. Still, payment for the change should be in full (less retainage) when the change is complete.

A word of caution on progress payments. Any payment schedule that trails expenses will force a contractor to work elsewhere to generate cash flow. That's a huge distraction you don't need. A contractor strangled for cash isn't going to respond when needed — making the CM contractor look foolish to the owner. Payment policy requires precise balance: Keep enough cash flowing to the contractor to keep him focused on your project. And hold back enough cash as retention to protect the owner's interest.

When progress payments are due on a calendar day, the contract will specify when invoices have to be submitted, how many days can pass before the invoice is either approved or rejected, and when payment is due. That resolves the "when?" issue. You still have to identify how much is due on each pay date. In a cost-plus (material and labor) contract, the amount due with each progress payment will be the sum of invoices received during the last pay period, plus the contractor's fee. For a fixed price contract, you need some other way to identify the amount of each progress payment.

The Schedule of Values

On a small job to be completed in several months, it's probably enough to have the contractor base the invoice amount on an estimate of the percent complete. For example, an estimate that the job is 50 percent complete would warrant issuing an invoice for 50 percent of the contract price, less any retainage. This percentage is subject to negotiation. Just check to be sure that work remaining is roughly proportionate to the amount not yet paid on the contract.

On larger jobs to be completed over a longer period, payment will be based on a schedule of values. To calculate the amount due at each progress payment, multiply the units installed during the pay period by the cost per unit in the schedule of values. A schedule of values can be used when the contract price is either a fixed sum or a cost-plus fee with a guaranteed maximum price (GMP).

If progress payments will be based on a schedule of values, that schedule should be approved before the first pay date. The contractor's estimate is usually a good start in developing a schedule of values. The schedule should provide a detailed breakdown of material, equipment and labor costs for each distinct part of the job. The total of the schedule of values should equal the contract price. That means every job cost has to appear somewhere on the schedule.

> **Approving the Schedule of Values**
>
> On a larger project, approval of the schedule of values marks a major milestone. Here's the best way to reach that milestone.
>
> - On bid day, get the contractor to commit to delivering a schedule of values within seven days. Never start construction without that schedule. Any leverage you have with a contractor is based on an approved schedule of values. If you have no schedule, you have no leverage.
>
> - Send the contractor's draft schedule of values to all concerned: the owner, designers, inspector, etc. Ask for comments and corrections. Insist on a response within one week.
>
> - During that week, crawl through the schedule yourself. Look for anything that's even slightly out of skew. For example, convert the labor cost for the month into work days for the month. Is that number reasonable? Or is it preposterous? If it doesn't feel right, dig into it nail by nail.
>
> - After one week, meet with the designers, the owner and inspector. Compare notes. Develop a consensus on what's right and what's wrong with the draft schedule of values.
>
> - Meet with the contractor to hammer out the final schedule of values. Use the consensus you've developed to reduce inflated values. You're done when everyone consulted can support the draft schedule.
>
> - I call the result the *God* schedule of values. From then on, every cost is either up or down from the God schedule, including change orders and credits.

For example, if the contractor's estimate for block masonry is $50,000 and if the job requires 5,000 square feet of block masonry, the schedule of values might show a cost of $10 per square foot of block, including labor, material, equipment and markup. Laying up 2,500 square feet of block during the pay period would earn a progress payment of $25,000.

The example for block masonry is the easiest case. The contractor may not have detailed labor and material costs for work done by subcontractors. Smaller subcontracts can be incorporated into the schedule of values as lump sum items, with no detailed breakdown. Request detailed material, labor and equipment costs for larger subcontracts that require more than a month to complete — especially if the subcontractor will receive progress payments.

Generally, you won't have to count the block or cubic yards or square feet of anything. If cost categories in the schedule of values are small enough, most work will be completed during a single pay period. That simplifies the task, to both your benefit and the benefit of the contractor.

All amounts on the schedule of values should include overhead and profit for both the contractor and subcontractors. I call that the *all-in number.* Insurance and bonds should be listed as separate line items. Some jobs have heavy startup costs, such as insurance and bonds, that have to be paid before work starts. In that case, you may want a separate schedule of values covering project startup.

If the contract includes allowances, alternate bids or unit pricing for some work, list those separately on the schedule and exclude those costs from the total contract price.

When reviewing the draft schedule of values, be alert for inflated costs early in the job. Front loading helps the contractor's cash flow but also reduces the incentive to complete later phases on schedule. Ask the contractor to document any costs that seem inflated and then submit a revised schedule. If inflated costs appear in a subcontract, get more information from the subcontractor. Breaking costs down into material, labor and equipment components should disclose any error or exaggeration.

Ask for a revised schedule any time approved changes in the job require a major change in the contract price.

A schedule of values is like a construction schedule. It's the contractor's opinion. Unless you see some cost that's well out of line, don't try to substitute your opinion for the contractor's opinion. Preparing a schedule of values should help the contractor find potential problems early in the construction process. That could be the primary benefit, not the schedule itself.

When payments are based on a schedule of values, the invoice has to include a lot of detail. You should see a detailed list of work done and materials delivered during the pay period for each item in the schedule. To avoid overcharges and to help plan cash flow, you'll need a record by line item for:

1. charges for work this pay period
2. what's been paid already
3. the amount of retainage, if any
4. change orders approved
5. the contract price (with changes)
6. the value of work yet to complete

We'll recommend an accounting system later in this chapter.

Final Payment

The contractor will submit an application for final payment when the job is substantially complete. Final payment won't be due until the building inspector has signed off on the final inspection and issued a certificate of occupancy. Work isn't done until the utility companies (water, electric, gas, cable) have authorization to begin service. When you receive the application for final payment, schedule a walk-through to identify work yet to be completed.

"Collecting the final payment sometimes requires unusual methods."

The walk-through should include the contractor, designers, inspectors and owners. Invite everyone to start their own punch list. When done with the walk-through, combine all lists into a single list. From that list, eliminate anything that's not required by the plans, specs or the contract. Note any missing items, such as hardware (keys), scratched or defective fixtures, and painter's holidays (missed spots). Be sure plumbing, HVAC and electrical systems work as expected.

If you note a serious defect, reschedule the walk-through for a later date. If only minor defects are noted, enter those defects on the punch list and recommend acknowledging substantial completion. If serious defects can't be fixed easily (such as a chip in a granite countertop), ask if the contractor is willing to issue a credit on the contract price. On a larger job, the contractor may offer to "bond off" some defect in return for acknowledgement of substantial completion. In essence, the contractor buys a bond (from a bonding company) equal to perhaps 125 percent of the estimated cost of repairs. The contractor either makes the repairs, or the owner collects on the bond. Others will finish the work.

When the contractor claims that punch-list items have been corrected, re-check the job. If you agree that every item on the punch list has been worked off, certify to the owner that it's time to consider making final payment. From that point, the contractor's only remaining obligation will be for any callbacks and warranty items.

Note one subtle point about final payment. The payment is due when the owner acknowledges *substantial completion,* not final completion. Substantial completion is when the project is ready for the intended use — with only minor punch-list items yet to complete. Final completion marks the end of the construction project — all punch-list items and all potential claims have been resolved.

Release of Claims

Final completion should be the end of the construction project. Except for warranty claims and obligations imposed by state law, final completion ends the contractor's obligation. The owner and contractor may want to sign a reciprocal release of claims, either conditional or unconditional. A conditional (limited) release doesn't bar claims discovered after final payment and excludes claims based on fraud or misconduct. The sidebar has a sample unconditional release of claims by the contractor.

> **Unconditional Release of Claims**
>
> Final completion should end work on the project. A release such as this puts an end to nearly all contract obligations. To be fair, you need two releases. In the first, the contractor releases the owner, as in the document below. The owner should also release the contractor in a separate but similar paragraph.
>
> *The acceptance of final payment by contractor constitutes a complete and unconditional waiver and release of any and all claims by contractor of whatever nature, and regardless of whether they are then known or unknown, and a complete and unconditional release of owner, and every person for whom owner is responsible, for any and all matters related to the contract or otherwise, except those claims which have been made in writing and identified by contractor as not having been settled at that time.*

Before making final payment, be sure that:

- All lien claims have been satisfied — ask for conditional waivers of lien from all contractors, subcontractors and suppliers. Conditional waivers of lien become unconditional when the owner's check clears the bank. We'll cover release of liens later in this chapter.
- As-built drawings, certificates, warranties and job records required by the contract have been delivered by the contractor.
- Punch-list items have been satisfactorily completed. With the owner's approval, accept a certified check or bond for an amount slightly more than the value of uncompleted work.
- The contractor has surrendered all master keys and keying schedules.
- Required certificates of insurance have been delivered. Any policy on the completed work issued to the contractor should be assigned to the owner when work is completed.
- The lender and all sureties that issued bonds have been notified and have given consent to final payment.
- If required in your state, a notice of final completion has been recorded. This starts the statutory period allowed for filing lien claims.

Release of Retainage

Retainage is money temporarily held back or retained by the owner pending completion of the project. Retainage is an incentive for the contractor to complete work to the owner's satisfaction and in good time. Many states set

> **New York Construction Contracts Act**
> **General Business Law § 756 to 758**
>
> The law covers most non-residential jobs valued at $150,000 or more, and residential jobs of more than three units, or 4,500 square feet.
>
> ➢ Unless the contract provides some other term, the billing cycle is a calendar month.
>
> ➢ The owner must approve or reject an invoice within 12 business days.
>
> ➢ Payment of undisputed amounts is required within 30 days after approval.
>
> ➢ There are only six legal grounds for rejecting a charge on an invoice.
>
> ➢ After giving 10-day notice, a contractor may stop work if not paid on time.
>
> ➢ The contract completion date is extended during any work stoppage due to delayed payment.
>
> ➢ Unless the contract provides some other term, <u>retainage</u> has to be released 30 days after final approval.
>
> ➢ Unless the contract sets a higher rate, the interest rate on late payments is 1 percent a month or fraction of a month.
>
> ➢ No matter what the contract says, disputes about compliance with New York law have to be resolved by expedited binding arbitration.

limits on retainage for some types of work (usually public works projects and non-residential private construction). Your state may require release of retainage within a few weeks or a month after final completion. The amount retained and the release date will be in the contract. Retainage is usually a percentage of each payment due.

Retainage can be a burden for contractors. It's considered good practice to limit retainage to some reasonable amount or period, such as by reducing the percent retained during later stages of the project. Once early subcontractors (such as the dirt contractor) have completed their work, it makes sense to release any retainage withheld on their work. Early release of retainage is considered good practice, but should always be at the discretion of the owner.

Retainage is usually considered the property of the contractor even while held by the owner. Some states require that retainage be deposited in a separate account pending payment to the contractor. Interest earned on retained funds usually accrues to the contractor. Where a separate retainage account isn't required, the owner could use retained funds for other purposes. For example, retainage could be used to repair defects or satisfy lien claims on the job.

Other Retainage Issues

Unless the contract provides otherwise, a contractor could withhold retainage on subcontracts even if there's no retainage on the prime contract. As the CM contractor, you should recommend that retainage on subcontracts be no more than retainage on the prime contract. If retainage is released early to the prime contractor, recommend that subcontractors receive the same consideration.

The purpose of retainage is to encourage faithful performance of the contract. There's no need for retainage if the contractor agrees to a deposit of marketable securities or a certificate of deposit or a letter of credit equivalent to what would be retained. If you accept a deposit in lieu of retainage, be sure the owner is authorized to liquidate that deposit to cure a default.

Interest on Past Due Balances

If the contract sets an interest rate on past due balances, that rate will be enforced by the courts. If the contract doesn't set an interest rate, the state's default legal rate will apply. But many states now require prompt payment on construction projects, especially public works projects and larger commercial projects. New York's law is a good example. See the sidebar on the previous page.

Some large public institutions simply can't pay bills in 30 days. For example, some payments by public organizations have to be approved by a board of supervisors. If the board meets only every other month, getting payment in 30 days may be problematic. If state law doesn't require payment in 30 days, and if the contractor doesn't object, build a 60-day or 90-day grace period into the contract. During that grace period, no interest accrues.

Materials Stored Off-Site

Progress payments will usually include charges for materials delivered but not yet used in the work. On some jobs, there may be legitimate reasons to order materials well in advance of actual need, and to store those materials temporarily somewhere other than at the site. Obviously, there's risk in storing materials off-site. If the contractor wants to store materials off-site and include the cost of those materials in progress payments, set some reasonable limits.

- ➢ Require prior approval. Written approval is best.
- ➢ The contractor should provide proof of insurance.
- ➢ Get a monthly inventory of what's stored off-site.
- ➢ Insist that off-site storage be marked with the name of the project and the owner.
- ➢ Ask to visit the storage site. Be sure the materials are secure.
- ➢ Get good documentation on the cost of stored materials.

- Get a pledge that stored materials won't be used for other purposes.
- Set a limit on the value of materials stored off-site.
- The risk of loss should be on the contractor until materials are installed on-site.
- Be sure that what's stored off-site is really suitable for the project.

Before recommending approval of payment for materials stored off-site, check the policy of the owner. Payment for materials stored off-site is prohibited on some public projects.

"We have a few extra charges here. I know you folks won't mind."

Pay or Reject?

The heart of a CM contractor's responsibility is recommending either payment or rejection of invoices. Recommending payment is easy. Just pass the invoice along to the owner. Attach to that invoice:

- a note that identifies the payment due date
- a recommendation on the amount to deduct for retainage, if any
- a note on any pending penalties or backcharges

Attach your updated spreadsheet detailing amounts paid to date and the amount remaining unpaid. If payment is based on a schedule of values, attach your worksheet showing how the amount due was calculated. We'll have more to say about creating this spreadsheet later in the chapter.

Certification of a payment request doesn't necessary mean that you've verified every charge. For example, it's not necessary to check every material requisition, examine the invoice from every subcontractor or count every spade of soil turned. Far better and easier, require a signed statement of accuracy on the invoice:

Contractor certifies that charges in this invoice are a correct, true and accurate statement of the amount earned and payable under the terms of this contract for the current payment period, and include no items for which payment has been made previously.

Omission of the signed statement of accuracy is grounds for rejecting payment. Of course, a signed statement doesn't relieve a CM contractor from the obligation of checking for errors. But it shifts the burden for uncovering errors. Until you discover something wrong in an invoice, you're justified in trusting the contractor.

> **Documenting Grounds for Rejection**
>
> A few years back I worked for 18 months as the supervising project architect for the Aladdin Hotel & Casino in Las Vegas. The project was both behind schedule and grossly over budget. The contractor was desperate to get the building up. By the time work was half done, the owner had already issued over 600 change orders. I had a doublewide trailer full of designers trying desperately to keep accurate information flowing to the contractor. Every hour or two, the owner's representative or the contractor would come busting into the trailer with a request for a change order or a request for more information.
>
> The firm I worked for had long ago burned through their design budget. But the owner insisted that the design staff keep working anyway — it was a big problem in the making. Just to be ready for the fight, I set up a notebook entitled "Unauthorized Work." Every time we got a request for design work, I entered in my book a job number, the date, time to complete the work, description of the work and who asked for it. My book would become the basis of our payment claim sometime in the future. In the end, the owner paid us based on that book.
>
> The next thing I know, the owner is making serious financial claims against the contractor. Soon the contractor was making similar claims against the owner. Before too long, lawyers are in our trailer going through our file cabinets. This went on for three months, right in the middle of our work. There were lawyers for the contractor and lawyers for the owner, all crawling through our files looking for documents that would support their money claims.
>
> Before too long, the main conference room of our trailer city was buzzing with high level, all-day meetings. Then, suddenly, one day, it all stopped. The "suits" reached a global financial settlement: The contractor would finish the job by a specific date or face liquidated damages of $1,000,000 a day. The owners would pony up something north of a $100,000,000 to satisfy existing claims. The rest of us just got back to work.
>
> That was probably the best possible settlement. Resolving 600 rejected claims one at a time would have required more effort than erecting the building.

Always check the invoice for math errors. If the contractor claims that 2,500 block have been laid during the pay period, the charge for that line item should be 2,500 times the cost per block. Then check to be sure the sum of all line items equals the invoice total. That's easy if you enter the invoice on a spreadsheet, as recommended later in this chapter. If you find a math error, do the intelligent thing: Correct the error on the invoice, approve the corrected amount and send a copy of the correction to the contractor.

Rejections

Rejecting an invoice takes more work. The contractor has to be informed of the reasons for rejection and should have an opportunity to respond before the owner is brought into the conversation. It's possible that you don't understand some item on the invoice or something at the jobsite. In some cases, there may be a technical defect in the invoice. For example, a lien waiver may be missing. In situations like that, it's best to give the contractor an opportunity to correct the invoice form and resubmit without sending a written notice of rejection.

Some states set procedures for approval and rejection of invoices for some types of work. New York is an example. There are only six statutory grounds for rejecting an invoice. But one of the six is failure to comply with the contract. Compliance with the contract can be a subtle issue subject to interpretation. In any case, undisputed portions of any invoice should be paid promptly. Even if your state doesn't have a prompt payment act, it's good practice to observe the following points when reviewing invoices:

- *Invoices are like money.* Give invoices the priority they would carry if they were your money. Review the invoice promptly and carefully. Recommend approval or rejection within a day or two.
- *Approve as much as you can.* Don't deny partial payment just because the invoice can't be paid in full. Public agencies are famous for withholding payment on a charge for $100,000+ because they can't approve a charge for $10-worth of nails. I know a CM contractor who's made good money sorting out issues like this.
- *Keep the contractor informed.* Ask for another invoice if you see a technical defect. If you recommend rejection of any portion of an invoice, notify the contractor or supplier promptly and in writing. Email is probably best. Advise the contractor that you'll reconsider if the defect can be remedied before the payment due date.
- *Explain rejections fully and in writing.* Describe exactly what's required to qualify for payment. If your rejection is based on something in the contract, cite the section and paragraph. If grounds for rejection are limited by state law, cite the specific code section that gives you the right to decline payment.
- *Keep the owner informed.* Write a full explanation. Cite the authority for the rejection. Advise the owner that the contractor, subcontractor or supplier involved has been informed of the grounds for rejection.

Grounds for Rejection

To avoid rejected invoices, walk the project with the contractor a day or so before each invoice is submitted. Discuss what's going to be on the invoice and what shouldn't be on the invoice. No contractor wants to be surprised by a rejected invoice. Hash out all likely disputes before the invoice is prepared.

All of the following are legitimate grounds for rejection. When rejecting any invoice or any item on an invoice, cite one or more of these grounds in your written notice of rejection. To make rejection easy, have a rubber stamp made up with the standard reasons for rejection. Then tick off the reason that applies. Be sure the rubber stamp includes a line for your signature, and the date.

Work is defective. Everyone can agree that no payment is due on defective work. The dispute will be over what constitutes a defect. Clearly, construction that doesn't pass inspection is defective. And anything that doesn't comply with the plans, specs or other contract documents is defective. Anything the owner considers objectionable is a candidate for rejection. Most disputes over defective work will involve workmanship: What is professional quality work? Get an opinion from the designer or engineer if you need help. If you feel the work doesn't meet professional standards and if others in the trade would agree, it's probably substandard. So no payment is due. If you've cautioned the contractor about substandard work and if the contractor's performance remains substandard, the last resort is to decline payment.

Claims filed against the contractor. Owners are entitled to withhold payment when a legal claim filed against the contractor relates to the job and could result in a claim against the owner. If the claim would be covered by insurance or a bond, the claim probably isn't a valid reason to withhold payment. Notice that we're talking about the *filing* of a claim. A mere threat to file a claim may not be grounds to withhold payment.

Failure to meet financial obligations. Owners are entitled to withhold payments due a contractor who isn't meeting obligations to subcontractors, tradesmen, material suppliers, taxing authorities or union welfare trusts. Again, any obligation covered by a bond or other guarantee isn't a valid reason to withhold payment.

Project can't be completed for the unpaid balance. You can reject an invoice when it's obvious that work can't be completed for the remaining unpaid balance of the contract. Again, this isn't a valid reason to withhold payment if there's a completion bond on the job.

Damage caused by the contractor. The owner can withhold from payments enough to cover a loss to the owner caused by the contractor. For example: damage done by the contractor to landscaping, a driveway, or an existing structure on-site. Again, this isn't a valid reason to withhold payment if the damage is covered by insurance.

Liquidated damages exceed the amount due. Liquidated damages are set by some contracts to compensate an owner for late completion. When liquidated damages due on completion are likely to exceed the amount due on the contract, the owner has the right to stop making payments. If the contract doesn't include a provision for liquidated damages, the owner may still collect damages for late completion. But those damages have to be established by a court or arbitration panel. Until established, the owner has no right to make a deduction from current payments due. Note that *force majeure* (good old Mother Nature) can provide valid grounds for extension of the completion date.

Persistent failure to comply with the contract. When relations with a contractor reach this point, there may be grounds to terminate the contract. If you stop making payments, the contractor will probably claim breach of contract.

> **Damage Caused by a Contractor**
>
> A little carelessness on a construction site can go a long way. Here are some examples:
>
> Not too long ago, I had a dirt contractor run into some underground asbestos pipe while grading the site. Work continued for several hours, spreading asbestos fibers throughout the site and into HVAC duct in adjacent buildings. Everything had to be cleaned up by guys in space suits — at a cost of three weeks and $30,000.
>
> Another time, a contractor didn't bother to tighten attic and roof drain pipe to the pipe flanges. A rainstorm opened up the flanges, flooding the ceiling, files, desks and computers in athletic offices at two school sites. I spent several miserable days getting the contractor to clean everything up or replace it. I presided at more than a few tense meetings between the owner and the contractor. Thirty days and $70,000 later, we were finally back to where we started.

At that point, the dispute becomes an issue of fact. Who was the first to breach the contract? An owner who stops payment for failure to comply with the contract can expect a counter-claim for breach of contract. The designer and insurance companies will be named as additional defendants. That's what I call the "money-go-round."

Failure to comply with the law. This is another issue of fact. When failure to comply with the law affects the entire job, you may have no choice but to withhold payment. An extreme example: The contractor insists on using unlicensed crews or refuses to take out a building permit. Occasional citation for code violations wouldn't be a reason to withhold payment as long as the contractor makes corrections. The contractor will always be liable for permit penalties and the cost of re-inspection after a failed inspection.

Failure to keep work progress on schedule. This is another issue of fact. Many jobs are completed a month or two behind schedule. That's not going to be a reason to withhold payment. But slippage in the schedule that approaches abandonment of the job would justify withholding payment — and probably be a reason to terminate the contract for cause. We'll discuss termination for cause in Chapter 13. This is a difficult matter to prove and will usually require the opinion of an outside expert, at considerable expense, even if the case is settled before trial.

Money owed on another contract. Banks have a right of offset — the right to transfer money from one account to meet an obligation of the same customer in another account. Owners will have the same right only if the contract allows offsets against an obligation of the contractor on a different job.

Discovery of an earlier defect. If you discover a defect in work already paid for or an error in an earlier payment, you're entitled to withhold from the next payment due. Payment of an invoice doesn't mean that work is permanently approved. All work is subject to review until final completion.

Backcharges. Some contracts give an owner the right to charge a contractor for neglect of contract obligations. The most common of these is failure to keep the site clean. The contract may allow an owner to clean the site and deduct the cost from the next payment. Before exercising the right to backcharge a contractor, send a written reminder, including a quotation of the relevant section of the contract. Backcharges shouldn't be a surprise.

When deciding to approve or reject items on an invoice, remember that wrongful rejection of an invoice may give the contractor certain rights. For example, wrongful delay may give a contractor the right to stop work or even claim termination of the contract. If a court or arbitrator decides that a payment was delayed without cause, the court or arbitrator may award both interest and attorney fees to the contractor.

Liens and Waivers

All states give lien rights to builders. A construction contractor, subcontractor, tradesman or material supplier who isn't paid can record a lien on the property being improved. This mechanic's lien becomes a security, like a mortgage, once perfected by a favorable court judgment. Property owners run the risk that payments made to a contractor, for example, could be diverted to some purpose other than paying subs and material suppliers. That would leave subs and material suppliers with lien rights against the property even though the owner has paid the prime contractor in full. The owner might have to pay twice for some materials and subcontract work.

Owners protect themselves from liens by getting a conditional waiver of lien in exchange for payment. A lien waiver by a contractor, subcontractor or material supplier is acknowledgment of payment. The conditional waiver of lien becomes unconditional when the owner's check clears the bank. If the check doesn't clear, the lien remains in effect.

To protect the owner from liens, require waivers of lien with each invoice. The contractor, each material supplier and each subcontractor who acquired lien rights during the pay period should waive those rights, confirming payment for all work and materials covered by that invoice. A conditional waiver of lien becomes unconditional (absolute) when the owner's payment is distributed to each lien claimant. Material suppliers and subcontractors covered in a prior invoice should confirm that previous conditional waivers of lien have become unconditional.

Lien waivers aren't required:

> ➢ On public works projects (federal, state or municipal). Instead, the federal Miller Act (40 USC 270) and state equivalents (Little Miller Acts) require payment bonds for protection of subcontractors and suppliers.

- If the contractor waives all lien rights in the contract. That's called a *no-lien contract*. Most states won't enforce a no-lien clause in a contract because doing that would defeat the purpose of the lien statutes.
- When the contractor has furnished a bond guaranteeing payment of all suppliers and subcontractors.

Even if you don't require lien waivers with each progress payment, get lien waivers before making the final payment. Lien waivers at project completion will extinguish all lien rights on the project.

"I'm not exactly sure what you mean by lean wavers."

Payment of Subs and Suppliers

The owner isn't responsible for distributing payments to subs and suppliers. That's a task for prime contractors. But you have an interest in seeing that subs and suppliers are paid promptly. That helps avoids lien claims and keeps the job moving on schedule. If you're concerned that subs and suppliers aren't being paid on time:

- Have the contractor supply a list of all subcontractors, material suppliers and tradesmen who could claim lien rights. The list should include an estimate of the value of the work or materials each will supply on the job.
- Notify those subs and suppliers each time the prime contractor's invoice is paid. Subs and suppliers will take it from there.
- Consider paying subs and suppliers directly. Remember, failure of a contractor to meet financial obligations gives an owner the right to withhold payment to that contractor — and pay the subs directly.
- Pay with joint checks — made payable to both the contractor and the subcontractor. The contractor can't negotiate the check without the sub's signature. It isn't a joint check if the named payee is the contractor *or* the subcontractor. That allows either party to cash the check and defeats the purpose of listing both as payee.

"Pay-When-Paid"

Subs and material suppliers expect to be paid for work completed, regardless of whether or not the prime contractor's been paid. That's the law in most states. Prime contractors are liable for payment of subs and suppliers even if they haven't been paid by the owner.

General contractors prefer to make their obligation to pay subcontractors contingent on receipt of payment by the owner. A few states permit prime contractors to make payment of subs contingent on payment by the owner. A "pay-when-paid" clause in subcontracts could relieve the contractor of any obligation to pay subs until the prime contractor has been paid. The effect of a "pay-when-paid" clause may be to deprive a subcontractor of both the right to collect under the contract and statutory lien rights. Mechanic's liens don't attach until payment is due. If the owner becomes insolvent, the subcontractor might never collect. That's why courts in most states won't enforce a pay-when-paid clause in a subcontract.

Keeping Track of Payments

Whether the job is done at a fixed price, for the cost of labor and material, or some combination of these (such as cost-plus with a guaranteed maximum price), someone on the owner's side of the construction team has to keep track of payments. Obviously, the contractor will be tracking charges and receivables. But someone has to be sure the contractor is minding the store. That's the job of the CM contractor. And it's a simple task if the job has:

- no change orders
- only one or two progress payments
- no rejected invoices
- no contract allowances, unit prices or alternate prices
- no liquidated damages
- no chargebacks
- no retainage

If that's the case, you'll need only a pencil and a scratch pad to track payments. But a CM contractor may never see a job like that. Most jobs are exactly the opposite:

- many change orders
- many progress payments
- invoice items are rejected
- the contract includes allowances, unit prices and alternate prices
- liquidated damages are set
- retainage is withheld

If you have jobs like that, an accounting system is required.

In many construction offices, the accounting department keeps records on payables, receivables and payroll, probably with one of the popular accounting programs. This manual isn't a guide to construction cost accounting. But you don't need to be an accounting whiz to record what most CM contractors need to know. You probably don't even need an accounting program. A modern spreadsheet such as MS Excel should provide all the computing power required.

Bookkeeping for cost-plus (time and materials) contracts is pretty simple. Fixed price contracts and guaranteed maximum price *(contractor at risk)* contracts will be more difficult. We'll start with the easy cases.

Time and Material Contracts

The owner pays subcontractors and material suppliers. Little bookkeeping is required because there's no maximum price. Use a spreadsheet to keep a list of payments by date. You'll need a running total by payment category: materials, services, subcontracts and fees. If your fee is a percentage of construction cost or if there's a guaranteed maximum price, the running total will be important.

Check the accuracy of invoices before recommending payment. Invoices from subcontractors should show the job address, the work done, and the dates when work was done. Check the math for errors. Charges for materials should include a detailed materials list, the date of delivery, and the delivery address. A receipt for materials picked up at a dealer will show the date, a list of materials and an account or charge card number. If asked, the materials supplier should be able to show a signature of the person who accepted delivery.

It's OK to reimburse someone for materials they bought for the job. But it's better to pay suppliers directly. That removes any doubt that a reimbursement might be income to the recipient. No Federal form 1099 will be required. But if you reimburse someone for job expenses, be sure the owner's file of reimbursed charges includes the original cash register receipt and the request for reimbursement.

Change orders, rejected invoices, unit prices and liquidated damages aren't an issue on time and material contracts. The owner has agreed to pay all legitimate charges, regardless of the source. Retainage may be an issue. That's one reason why you need separate totals for subcontract and material costs. If you're withholding 5 percent on subcontract costs, for example, you need to break out subcontract costs and keep a running total of amounts retained.

List invoices by date in rows going down the spreadsheet. Your spreadsheet will grow longer with each progress payment. To the right of each invoice, show the amount of the charge. Insert a total at the bottom of the page.

On large public projects, inspectors and auditors from the public agency may be required to monitor labor, material and equipment records *(called a certified payroll)*. An inspector or an auditor may have to sign off on every timecard on the job. If labor, material and equipment expenses are monitored by an auditor or inspector, attach those records to the back of monthly invoices.

Fixed Cost Contracts

If payment is by job phase, keep a spreadsheet showing the date of each invoice and the job phase. In a column to the right of each listed invoice, show the amount. When a new invoice arrives, insert new rows showing the date, job phase, and amount. At the top of the *Amount* column, enter the contract price. At the bottom of the column, calculate the amount still due (contract price minus invoices approved equals the amount due). Every payment approved should reduce the remainder due on the contract price. Add a comment if an invoice is rejected. The spreadsheet will grow longer (more rows) with each additional payment.

When there's a change to the contract price, copy the entire *Amount* column. Then paste that column one column to the right. Change the old *Amount* column to hidden text. Change the contract price at the top of the new *Amount* column to the new contract price. The *Still due* figure at the bottom of the *Amount* column should have the correct figure based on the new contract price. Use the Insert Comment feature to document each change: The item changed, the date approved, and the amount of the change.

Treat contract allowances, unit prices, alternate prices and liquidated damages the same way you treat change orders. Copy the old *Amount* column to a new column to the right and change the old *Amount* column to hidden text. Then enter the new contract price at the top of the column. If the job includes retainage, add a column that accumulates amounts retained.

Schedule of Value Contracts

When payments are based on a schedule of values, each invoice will include a lot of detail. You should see a list of work done and materials delivered during the pay period for each item in the schedule. To avoid overcharges and to help plan cash flow, you'll need a record by line item for:

1. the contract price (with changes)
2. charges for work in each pay period
3. the total of invoices approved to date
4. the amount of retainage, if any
5. the value of work yet to complete

Start your spreadsheet with a breakdown showing items on the schedule of values. The contractor can probably supply this list of items and values in a format you can copy and paste into a blank spreadsheet. Head the first column *Items*. Above the list of values, add a column heading, *Total Cost*.

Add a new column to the spreadsheet for each progress payment, working left to right. Head each of these *Charges* columns with the invoice date. Opposite the correct category in the *Items* column, enter the amount invoiced. The sum of charges in all categories for that invoice date should be the same as the invoice total.

One column to the right, head a new column *Still due*. Enter in this column a formula which accumulates figures in the *Charges* columns for that row and subtracts all charges from the figure in the *Total Cost* column for the same row.

Each time you receive a new invoice, insert a new *Charges* column to the right of the previous *Charges* column. Populate that column with charges listed on the invoice. You'll see at a glance:

> ➢ amounts invoiced to date for each value category
> ➢ the total of all invoices approved
> ➢ amount still due for each value category
> ➢ the total still due on completion

When there's a change to the contract price, copy the entire *Total cost* column. Paste the contents of that column one column to the right. Enter changed figures in the new *Total cost* column. Change the old *Total cost* column to hidden text. Be sure the *Still due* column references figures in the new *Total cost* column, not the old *Total cost* column.

Treat contract allowances, unit prices, alternate prices and liquidated damages the same way you treat a change to the contract price. If the job includes retainage, add a column that accumulates amounts retained. Make liberal use of the Insert Comment feature to explain rejected invoice items, backcharges, changes in the work and discrepancies that need explanation.

The system described isn't perfect and may need some tinkering to meet your requirements. But this system provides a good audit trail and doesn't require any specialized accounting software. Using only this spreadsheet, you should be able to retrieve enough information to answer nearly any question about payment.

Face Time

On a large project, you'll need an accountant to track expenses against budget. On a small job, keeping a spreadsheet will be enough. Regardless, deliver financials to the owner at least monthly. Weekly is better. This is your opportunity for face time with the owner. When you're with the owner regularly, it's pretty hard for others to deliver bad news first. That helps protect against anyone undermining your authority.

As a CM contractor, I make it a point to be in front of the owner at least once a week. Daily is even better. I deliver good news to an owner the day it happens. When there's bad news, especially financial bad news, I want the owner to hear it from me first in the same hour that I hear it. I *absolutely insist* that the designers and contractors notify me of problems no more than 5 minutes after discovery. If they've got a budget overrun, done some bad accounting or been embezzled by their superintendent, I want to know it *now, not later*. I never, never, ever want to hear about a problem from the owner... *never, ever!* That goes double when the problem involves money or budget.

The better prepared you are on financial issues, the greater your influence will be with the owner. Here's an example.

A few years ago, I was appointed to manage a project for a public agency — a school district. We were in the early stages of design. There was a rough budget and we had no complete and final commitment on funding. As much as $100,000,000 would be required. Designers and contractors had many opinions on what the project would cost; we were awash in economic opinions. As a CM contractor, that's not a good position to be in. You don't want to be a victim of a budgeting process that's run amuck. Better that you take the lead in developing a budget for the project.

In this case, I knew the rough budget number the school district wanted. So I went to my accountant and spent an hour analyzing costs to date. With those figures, I rushed off to meet with the designers and contractors. In a 3-hour meeting, I literally forced the design firm to solidify their concepts and the contractors to project their *"all in"* costs. At the end of the meeting, I had a believable overall project budget, knew my costs to date and had a very good idea what the design work and construction was going to cost. From that point on, any escalation in construction costs would have to be offset by a reduction in the design itself. I then made the contractor the designer's watchdog and the designers the guardians of the owner's program, a self-policing maneuver on my part. The designers couldn't add details to the project without the contractor's cost approval and the contractor couldn't eliminate pieces of the owner's program without the designer's blessing. When they both agreed, we built it on budget and on time.

No sooner had I finished that design/budget meeting than a couple of board members called me with a request that we meet in an hour and nail down the budget. When we met, I had all the facts on hand — by the skin of my teeth — and carried the day.

Up Next

Chapter 10 is the last of three "communication" chapters. In Chapter 8, we worked to keep the owner informed. In this chapter, we followed invoices and payments. In the next chapter, we're going to keep the trade contractors informed: Answering questions, handling objections, holding meetings, and avoiding disputes.

Chapter 10

Communicating with Contractors & Suppliers

THIS CHAPTER IS AN antidote to the poisonous misunderstandings described in Chapter 2, the case of Thurber Lumber Co. Inc. v. Marcario. If it's been a while since you read Chapter 2, or you didn't read that chapter carefully, this might be a good time to skim through those pages once again. Marcario made a $59,391.51 mistake. He left the supplier (Thurber) and the subcontractor (Nu Frame Contracting) in the dark regarding his role in the project.

It's easy to keep contractors and suppliers informed about your role. There shouldn't be any confusion:

> *When ordering supplies* for the owner, use the Purchase Order Cover Sheet shown in Chapter 6.

> *When requesting a quotation* for the owner, use the Request for Quotation Cover Sheet shown in Chapter 6.

> *When ordering supplies or closing a deal with any contractor or subcontractor,* be sure the owner signs the purchase order or agreement.

Also while in Chapter 2, take a look at the sidebar *Agency CM Contracts*. The subject of this book isn't a construction manager who acts as an agent of the property owner — has authority to spend the property owner's money. An agent of the owner and the owner are much the same for legal purposes. That's not what we're discussing here.

As used in this book, *CM contractor* refers to an independent consultant who makes recommendations and gives advice. As a consultant, you're insulated from the legal and financial responsibility carried by traditional prime contractors. We'll start this chapter by defining the limits of a CM contractor's authority: How close can you come to the role of a traditional general contractor and still preserve your status as a consultant? In practice, you can do more than simply give advice.

Limits of Authority

What we describe in this chapter is consistent with a CM contractor's role. Nothing we recommend in this chapter will make you either:

➤ an employee of the owner, subject to withholding and worker's comp requirements

➤ a prime contractor, financially and legally responsible for work done on the job

Obviously, the limit of your authority has to be settled one-on-one with the owner right up front — before work starts. For the purposes of this chapter, we're going to assume you've received either explicit or implied authority to handle each of the tasks and responsibilities listed below.

With the owner's approval, the authority of a CM contractor may include all of the following:

➤ assist in designing and bidding the project and reviewing insurance requirements

➤ visit the site and view the work at any time

➤ approve cosmetic and minor nonstructural changes in the work

➤ administer trade contracts through the general contractor

➤ check for compliance with the plans and specs, review substitutions and assist in permitting

➤ interpret and clarify the plans and specs, subject to the architect of record authority

➤ ensure compliance with law, ordinance and manufacturer's instructions

➤ evaluate the quality of workmanship, materials and manner of performance

➤ reject work which does not comply with the contract

➤ make judgments about compliance with the project schedule and budget

➤ report defective trade work to the general contractor

- preside at meetings of contractors and subcontractors, assuming the architect or general contractor isn't filling that role
- require that all communication intended for the owner be routed to you first
- suspend work to prevent performance of defective work, or for safety issues
- require additional inspection or testing regardless of the completion status
- collect documents which become the property of the owner
- approve or reject requests for payment, including final payment
- respond to requests for a change in the contract
- determine the dates of substantial completion and final completion
- resolve disputes between contractors, suppliers and government authority

With this authority comes some responsibilities. To avoid misunderstandings and disputes, you should:

- be fair and impartial; but remember that you represent the owner
- respond promptly to questions and requests from contractors and suppliers — preferably within 24 hours, but never later than 48 hours
- refer issues you can't resolve promptly to the designer or owner; make the referral the day you determine the issue has become unresolvable
- preserve a written record of your important decisions in your own legal file

If your authority includes all of the responsibilities just listed, you might wonder, "What's excluded?" The answer is "Plenty." Everything on the list below is strictly the responsibility of the owner. CM contractors *don't* have authority to:

- revoke, alter, relax, or waive any requirement in the plans or specs
- approve anything in conflict with law, ordinance or manufacturer's instructions
- accept any portion of the work without approval of the owner
- issue instructions contrary to what appears in the trade contracts
- stop work, except to avoid the performance of defective work, or for a safety violation
- terminate the contract

Because you don't control construction means, methods, techniques, sequences or procedures, you're not responsible for acts or omissions of trade contractors. Because you don't supervise the work, you're not liable for construction defects.

Also, understand that your decisions aren't binding if inconsistent with the plans, specs or the contract. No contractor or supplier will be bound by any instruction you give if that instruction is contrary to anything in the plans, specs or the contract. A contractor or supplier will always have the right to insist on referring a decision to the owner.

Finally, remember that you have authority on the jobsite only because you have the complete backing of the owner 24/7. The first time the owner publicly overrides one of your decisions on a design or construction process, your authority on-site gets discounted by 50 percent. If that happens five or six times, get another job, because you've lost all control of the process. If you have process management issues with the owner, settle them in private; job meetings aren't the correct forum for debates between you and the owner.

Previous chapters covered most items on the preceding CM authority list. Change orders, resolving disputes and project closeout are the subjects we'll cover in the last three chapters in this book. We'll use the remainder of this chapter to go into detail about other items on the list: manufacturer's instructions, interpreting the contract, interpreting the plans and specs, meetings, suspension and termination of the job for cause.

Manufacturer's Instructions

All construction material manufacturers want their materials used as widely as possible. But building inspectors and plans examiners are reluctant to approve use of new materials that haven't been tested. To win acceptance from building officials, manufacturers submit new products for testing by any of the dozens of standards development organizations (SDOs). When approved, the SDO publishes the new standard and describes the conditions under which tests were performed. Once the standard is published, use of that material will be approved routinely at your local building department office — so long as installation complies with standards used in testing. Installation standards are distributed with the material and should be made available to the inspector. Installation that doesn't comply with the published standard isn't going to be approved.

"The label on the can says to apply the paint in one coat. . . and this is the coat."

> **Flame-Spread Material Class**
>
> Pay close attention to the flame-spread rating of materials. Many will be referenced in the plans. Inorganic materials such as brick and tile are ASTM flame-spread Class I. Most whole wood materials are Class II. Plywood, particleboard and hardboard are usually Class III. The state fire marshal enforces limits on flame-spread materials used in construction. In Las Vegas, for example, the state fire marshal sets a limit of 10 percent Class II flame-spread material in commercial hotels. All those beautiful wood entrance porticos you see at Las Vegas casinos are actually Class I steel framing covered with stucco and hand painted to look like wood. The acres of wall coverings you see throughout the casinos are Class I rated materials applied over drywall and steel studs.

These national standards and tests have ASTM master data numbers. You'll see these ASTM master data numbers on plan sheets and in the specifications. Most building departments will have reference copies of key standards and tests on file.

Most mechanical and electrical fixtures come with instructions. Many other materials come with instructions: paints and coatings, windows and doors, adhesives, siding, roofing, and lots more. Think of these instructions as extensions of the building code. If you use the right material, but in the wrong way, you're probably in for a nasty surprise on inspection day.

Anyone who's been around construction for a while has seen plenty of installation instructions. For siding, instructions will describe how the product has to be supported and nailed. For roofing, instructions will cover spacing and roof pitch. For electrical lighting fixtures, instructions will identify minimum clearances from combustible materials.

There are at least two good reasons to require compliance with the manufacturer's installation instructions. First, the inspector won't approve installation that doesn't comply. Second, if the fixture or equipment comes with a warranty, faulty installation will void the warranty.

Verifying compliance with manufacturer's installation instructions is part of every CM contractor's responsibilities. Ask the installer for a copy of the instructions. Be sure installation complies. Retain those instructions for reference on inspection day. At project completion, transfer to the owner your file of all instructions and manufacturer reports collected during construction.

Interpreting the Contract

Contracts are written to prevent disputes. Good construction contracts narrow the room for dispute by clarifying what the parties intend. But construction is complex. It's nearly impossible to anticipate every possible dispute. If there's some ambiguity in the contract, a CM contractor is on solid ground when applying any of the following rules of contract interpretation:

> ➢ *Correct obvious errors.* Omission of words or phrases in a contract or obvious typographical errors shouldn't defeat the purpose of a contract. As long as the meaning can be inferred from a contract, the contract should be enforced as intended.

> *Terms should be given their commonly understood meaning.* Words used in the construction industry should be recognized in their construction industry context. Any word not defined in the contract and which doesn't have a well-known technical or construction industry meaning should be understood as defined in the usual dictionary definition.

> *No violation of law.* Nothing in the contract should be interpreted as requiring a violation of law or regulation.

> *Meaning of the words "will" or "shall."* These words mean compliance is required. The words "approved," "acceptable" and "satisfactory" mean "approved by," "acceptable to" or "satisfactory to" the owner unless otherwise expressly stated.

> *"Including" or "includes" are examples.* These words don't identify limitations unless preceded by non-limiting language such as "without limitation."

> *Organization of the specifications doesn't imply a division of the work.* Arrangement of plans or specs isn't intended to divide the work among subcontractors. Division of work among subcontractors is the responsibility of the prime contractor. Most specifications are now organized by CSI (Construction Specification Institute) number. All Sweet's catalogs are set up that way. The CSI format starts with site grading and proceeds in about the same order as construction: concrete, framing, utilities, roofing, siding and painting.

> *Published standards cited in the specs* refer to the latest edition, unless otherwise noted.

> *Waiver of contract* provisions shouldn't be assumed. Temporarily ignoring some provision in the contract shouldn't be interpreted as a waiver, or an unwritten revision to the contract.

The *Restatement (Second) of the Law of Contracts,* published by the American Law Institute, offers additional help when interpreting construction contracts. For example:

> Contract language has to be interpreted in the context of the circumstances that existed when the contract was signed. § 212

> Where two parties to a contract give different meanings to a contract term, the party who understood only one meaning will prevail if the other party understood both meanings. § 201

> If neither party understood what the other intended, or if both parties knew the other intended something else, there is no binding agreement on that issue. § 20

> If an essential term is omitted from an otherwise valid contract, a reasonable term will be read into the agreement. For example, omission of a completion date will require that the job be finished in a reasonable time. § 204

> Any oral agreement made during negotiation of a written agreement is void to the extent the oral agreement is inconsistent with the written agreement. § 213

> A CM contractor has the right to assume that construction prints will be interpreted as generally understood in the construction trade. For example, a reference to 2" x 6" lumber doesn't require use of lumber actually measuring 2" x 6". § 222

Interpreting the Plans and Specs

Plans and specs should identify the labor, material and equipment required to complete the project. The plans and specs are defective if a reasonably skilled construction contractor doing similar work in the community and following generally accepted trade practices couldn't use the plans and specs to identify the labor, material and equipment required. Any time the plans and specs are defective on some point, you're going to get a request for an interpretation.

Plans and specs are patchwork, usually prepared by copying and pasting from other sources. Mistakes are common. The most common mistakes will be about:

> what's included and what's excluded

> what the code requires

> an inconsistency, such as a conflict between the plans and specs

> something that can't possibly be, such as a room without a door

"Tutamuk, I don't care what my wife told you about feng shui; follow the plans!"

Some ambiguities can be resolved by convention. For example, plans and specifications seldom refer to items such as framing nails, scaffolding or slab forms. But nails, scaffolding and forms are rightly part of most jobs, even if not on the plans. Taken literally, nails, scaffolding and slab forms would be an extra charge on most jobs. However, we don't know any contractor who would make that claim.

Other ambiguities can be resolved by rules of interpretation, either narrow or broad. A narrow (literal) interpretation of what's included in the plans and specs generally favors the contractor. A broad interpretation of what's in the plans and specs generally favors the owner.

Narrow Interpretation

If the plans and specs don't identify or describe some detail or material, that detail or material will be part of the job only if generally accepted trade practice recognizes that detail or material as an essential element of good and skillful construction. Note that this standard is based on the opinion of other contractors (not architects, engineers or owners) in the community. So, even under a narrow interpretation of the plans, items such as framing nails, concrete forms and scaffolding would be considered part of the job. They're not on the plans, but nearly every contractor would consider nails, forms and scaffold to be required.

Broad Interpretation

Unless something is identified as furnished or installed by others, the contractor should plan to furnish, assemble, install, finish and connect each item in the plans and specs. Work not described in the plans and specs will be part of the job if it's:

> ➢ consistent with the purpose of the plans and specs,
>
> ➢ reasonably inferable from the plans and specs, and
>
> ➢ necessary to produce a complete building ready for the intended use.

> **Plans vs. the Specs**
>
> On larger jobs, you'll typically find a sentence in the architect's specifications that reads as follows:
>
> *If there is a conflict between the plans and the specifications, the specifications shall govern.*
>
> Here's the reasoning. Plans may take as much as a year to complete and may involve several design and engineering firms and a hundred people, all quite capable of making a mistake. Specifications, on the other hand, are prepared by one or two people as one of the last steps in the design process. If the plans and specs are inconsistent, it's probably the specs that are right.

Under a narrow interpretation, a home plan that doesn't show hot water piping is a plan to build a home without access to hot water. Under a broad interpretation, a contract based on the same home plan would be a contract to build a home with hot water piping to fixtures designed for the use of hot water.

Which interpretation applies on your job? As a representative of the owner, you're going to argue for a broad interpretation of the plans and specs. It helps if the contract includes language requiring a broad interpretation. But even if the contract doesn't suggest how the plans should be interpreted, it's fair to argue that the plans require quality work. Any ambiguity should be resolved in favor of quality construction.

Within the plans:

> ➢ dimensions written in numbers take precedence over scaled measurements
>
> ➢ notes and schedules take precedence over lines on the drawings
>
> ➢ large scale drawings take precedence over small scale drawings
>
> ➢ schedules take precedence over notes or other directions

Conflicts

What if the plans show studs 24 inches on center and the specs refer to studs 16 inches on center? If the contractor misses the reference to 16-inch-stud spacing in the specs, the job estimate and contract price will be based on drawings that show studs 24 inches on center. Suppose further that the building code requires studs 16 inches on center and the framing has already been completed with studs 24 inches on center. Who pays to reframe the job with studs at 16-inch centers — or to insert another stud between each 24-inch on-center stud, leaving walls framed at 12-inch centers and overly stout.

Both the CM contractor and the trade contractor will search for language in the contract that shifts responsibility to the other party. Who wins depends on whether the plans or the specs are given priority. To resolve this issue, the construction industry follows several conventions. These conventions are based on probability — the probability that someone truly considered the issue and made a decision that became the intention of the designer and owner.

For example, if the words *2" x 4" studs 16" O.C.* appear in a contract change order, it's almost certain that the issue of stud spacing was considered very carefully by all concerned. There's almost no possibility of a mistake. That's why change orders take precedence over every other document.

The set of priorities that follows is based on reasonable expectations.

Within the contract, plans and specs and incorporated standards:

- anything required by one contract document is required by all contract documents,
- the specifications generally take precedence over the plans,
- anything written in longhand takes precedence over anything printed. Approved changes take precedence over the original plans or specs,
- subsequent changes take precedence over prior changes,
- anything in the contract, plans or specs takes precedence over any manual, industry standard, recommendation, regulation, guidelines or instructions incorporated by reference into the contract. For example, the plans and specs would take precedence over a reference to the Western Woodworking Institute's means and methods of cabinetmaking specifications.

Within the specifications:

- product performance requirements take precedence over a named product or manufacturer,
- other clauses in the specs take precedence over anything incorporated by reference.

- specific notes take precedence over general notes
- bottom elevations of footings take precedence over any general notes

If required by the contract, inconsistencies can be resolved by giving precedence to the more restrictive, higher quality, more demanding provision in codes, safety orders, plans, specifications, referenced manufacturers' specifications, and industry standards.

Magic in the Building Process

I can't leave the subject of contract interpretation without making two points. The first is to draw a distinction between what you see on the plans and specs and what the designers intend. Contract interpretation may be useful in settling disputes or winning arguments. But don't take your eye off the big picture — which is, what the designers intend.

After many years of working with contracts, plans and specs, I'm convinced that the intent of the plans and specs is more important than the words and drawings. If you understand what the plans and specs mean, you can make great progress by following the designer's intent. I know that sounds like heresy, especially coming from a registered architect. I've heard building inspectors make the same point when interpreting the code. It's not the words of the code; it's what the code *intends* that's important. I feel the same is true when it comes to interpreting the plans and specs. I've seen a public sidewalk proposed to be built where a fire hydrant already existed, in direct conflict with code width requirements. I relocated the sidewalk around the hydrant, informed the architect who later did an as-built, and let the contractor proceed with the work. Another time, the drawings for a high-rise building showed an air conditioning duct running through the elevator shaft. I moved it and followed the architect's intent, not the words.

The second point is related, but slightly different. Plans alone can't run a construction project. During my time in the trenches, I came to understand that there are two forces at work throughout the design and construction phases. The first is the legal force — contracts, drawings, specifications, change orders and field orders. The second is the force of the building process itself.

Documents that make up the legal force are prepared by a host of highly-trained professionals dedicated to expressing their intent in words and drawings. Within the first few hours after work starts, a very large portion of that legal force becomes somewhat secondary. What remains and becomes primary is the intent. Words and drawings alone fail the building process almost immediately — they succumb to the speed, noise and thunder of getting on with the work. The legal force makes way for the building process through consensus that we're all headed in the same direction, towards the same goal. The authority of the written word is almost always supplemented, revised and augmented by the spoken word.

Inside the building process, verbal commitments often take precedence over the written words. Thousands of times, I've struck deals with designers or contractors to get something done right now, today. Only later did that verbal commitment morph into words on paper.

> **Performance Specs vs. Brand Names**
>
> Job specifications can describe materials by performance characteristics or by brand name.
>
> *Performance specs* identify material characteristics such as durability, workmanship, economy of operation or suitability for the intended purpose. A contractor can select equivalent material from any manufacturer or supplier.
>
> *Brand name specs* require use of the make and model number from a particular manufacturer. Any substitution has to be approved in advance.
>
> Generally, job specs for public works construction are performance specs. If a brand name appears in the specs, it's only for the sake of convenience. Any equivalent product from another manufacturer would be acceptable.

Every successful project is the product of harmonious relations. As a CM contractor, you need to keep and nurture consensus. If you can do that, the project will hum with excitement. Things will get built all over the site. But regress to enforcing exactly the written words (the legal force), and you risk sacrificing the building process. You won't be running the job any more. The legal force will be in charge. When a job is governed solely by words written on a page, expect progress to slow by 50 percent. As the CM, one of your main functions is to sweep the design and construction process clean of obstacles that would impede the progress of the project. Above all, keep the decisions flowing and the project moving!

Contracts are necessary. They define intent and settle disputes — in court, if necessary. But the building process requires more. The real magic lies in the ability of a CM contractor to orchestrate design, money, contracts, budgets, schedules, contractors, and even the owner. My advice is to take the fight out of getting a job built. Make it easy to design and easy to build. Replace the legal force with the force of common sense and common purpose. That's the best way I know to expedite any job, and the real art of building!

Correcting Design Defects

Ambiguities, inconsistencies and omissions in the plans and specs that can't be resolved by interpretation are considered design defects. Every designer can make a mistake, so it's reasonable to believe that every set of plans includes at least a few minor defects. Generally, design defects will be corrected at the expense of the owner. Assuming the contractor is acting in good faith, the contractor is entitled to rely on the work of a designer selected by the owner. But there are exceptions.

The trade contractor should bear the cost of correcting work that doesn't comply with the code if the contractor knew or reasonably should have known about a defect in the plans and didn't give notice. Some contracts make the contractor liable for the cost of correcting work that doesn't pass inspection if the contractor should have recognized the problem.

Most drawings include a statement requiring contractors to speak up about flaws discovered in the plans or specs:

The contractor shall notify the architect immediately before proceeding with the work.

It's not common, but it happens, and you should be aware of it — contractors have used plan defects to extract extra money from an owner. They build according to plan, and then:

1. discover the defect
2. report the defect to the owner
3. blame the architect
4. charge the owner to tear out what was built and rebuild it correctly

Incomplete Plans

I'll share an architect's secret with you. When an architect delivers a set of plans and says they're complete, they both are and they aren't complete. Plans are graphic representations of what's intended. Sometimes plans merely suggest. You have to infer many things. In that respect, plans have a lot in common with the cartoons you see in many chapters of this manual.

I once received a set of plans that depicted an existing power pole on-site. When I went out to visit the site, the power pole depicted on the engineer's plans was two city blocks over and three blocks down. Fortunately, I made that discovery with the owner standing right beside me. $100,000 later, we fixed that *design defect*.

> *"If the contractor gets stumped or the problem becomes structural or requires further engineering, he calls the architect."*

Plans for remodeling are notorious for being incomplete. What you get is 80 to 90 percent complete. Why? Because no owner wants to pay for a truly complete set of plans. That might cost as much as the building itself. So, to make the cost of plan preparation tolerable, designers draw most of the building and the most important details, and leave the rest of the design to the contractor — basically guessing and solving the unknowns as the work proceeds. If the contractor gets stumped or the problem becomes structural or requires further engineering, he calls the architect.

Here's a fairly common problem: The contractor tells the owner the plans aren't complete and are defective from front to back. The owner doesn't know any better and calls the architect, who tries to explain "cartooning" to an owner who's no longer sure he hired a first rate design professional. I've been through this scenario a thousand times. I advise the owner, in private, that the plans are seldom truly complete — 10 to 20 percent incomplete is normal. Then I tell the contractor to relax and proceed with the construction, and anything that's truly missing will be provided as we go along. Then we all go back to work.

The inherent danger lurking inside 80 percent complete plans is *never, ever* obvious to the owners — the budget is also short 20 percent. This is a classic bidding disaster in the making. The owner thinks the plans are complete, because the architect says they are, and sets the budget accordingly. The architect agrees to the budget and the project goes out to bid. But architects are terrible at cost estimating. Often this shortfall comes to light during bidding, when all the bids come in over budget, and then the project has to be redesigned and rebid.

If no one brings the shortfall to the owner's attention and the bid is awarded, the owner will find out eventually, when the change orders to cover the missing plan information start coming in. The amount of missing budget money typically turns out to be 15 to 20 percent and everyone is looking a little dumb, except the contractor who, of course, usually knew it all along. The only safe way to head off this kind of design budget catastrophe is to have the owner add 15 to 20 percent to their budget at the very beginning of the design process. Obviously this isn't a popular move, but you'll be a hero on the back end of the project, and the owner will be singing your praises to your next client.

Job Conferences

Any time more than two or three contractors or subcontractors will be working on a job simultaneously, it's good practice to have a conference. The jobsite is usually the best place for a meeting, and the best time is often a few minutes before the start of the work day. Not many contractors or subcontractors are eager to attend job meetings. Most would rather be making money — working with their tools. But you'll generally have perfect attendance if the contract assesses a nominal fine for failure to attend or send a representative to job conference.

The conference is a good place to introduce the people who'll be involved in the project and their chain of authority. You can also pass out your directory of names, addresses, telephone numbers and email addresses.

Good topics for a pre-construction conference include:

- the proposed construction schedule
- job procedures, such as how to get shop drawings, submittals or substitutions approved
- procedures for handling change orders
- construction site requirements, such as dust and erosion control, storm water management, project signs, cleanup and housekeeping, temporary facilities, utilities, security, parking and traffic
- safety procedures, including who is to have badges and keys, and screening for drugs and felons
- quality control, testing, notices and abatement issues
- inspection procedures
- any requirement for certified payrolls
- value engineering and buildability reviews
- the handling of payment requests
- temporary utilities, debris disposal, storage areas, potential conflicts among trades, compatibility problems, time schedules, weather limitations, space and access problems, structural limitations, as-built drawings, site protection, government regulations, environmental issues, and permitting

Each representative at the pre-construction conference *must* have the authority to make decisions for their trade on topics discussed. That's the reason each attendee is at the meeting. Conference attendees should include a representative from every major subcontractor and supplier. Always include a representative of the project designer.

Distribute an agenda when the meeting is announced. A day or two after the meeting, distribute a list of *takeaways* — action items that require a specific response. Send reminders if action items aren't completed as required.

Phone or web conferencing can be nearly as effective as an on-site meeting and will be appreciated by contractors based at more remote locations. Phone or web conferences work especially well for the last few meetings on most projects, when only a few issues remain on the agenda.

If you hate paperwork as much as I do, consider delegating the chairmanship of job conferences. For the first jobsite meeting, I usually have the prime contractor installed as chair. That relieves me of the job of prepping for the meeting and doing the meeting minutes later. In the design phase, if I have multiple sites, I put one contractor in charge of overseeing all of budget development for the meetings. An architect should prepare minutes of design meetings. I usually chair the design meetings. I have the prime contractor chair jobsite meetings because the contractor will be on-site most of the time. That's not true for either the architect or the CM contractor. Both the architect and the CM contractor will be playing catch-up with what's going on at the jobsite. In my opinion, putting the contractor in charge of jobsite meetings yields more accurate and more productive conferences.

Suspension of the Job

At the beginning of this chapter, we listed tasks that can fall within the authority of a CM contractor. One item on that list was *suspend the work*. Suspension of work by a CM contractor should be considered more an emergency measure than a policy decisions. For example, a CM contractor might suspend work if there's a serious safety violation or if work is being done that can't possibly be approved. If you recognize a serious problem that's going to result in a loss, give the word to stop work. This type of suspension is usually called *discretionary*. We're also going to discuss another type of suspension: suspension for cause. Discretionary suspension is within the authority of a CM contractor. Suspension for cause is not. Suspension for cause requires a decision by the owner.

Any time the job is suspended, the contractor has a potential claim for delay or grounds for a change order. So it's wise to be certain about why the job has to be stopped before ordering tradesmen to put their tools down. Any suspension that lasts an unreasonable time gives trade contractors grounds to terminate their contracts. The owner should consult legal counsel before stopping work for cause. It isn't something to do in a fit of pique. There's a real financial risk.

Discretionary suspension can be for any reason, but is usually the result of something unexpected — such as discovery of unsuitable soil, failure of financing, or the need to redesign some part of the job. But if the job requires a change, it's better and cheaper to stop work as soon as the facts are clear. An order to suspend work can be oral, but should be followed by prompt written confirmation to the owner, the contractor, and each trade contractor involved.

Suspension for Cause

Suspension for cause is like an administrative time out. The owner feels a trade contractor is making a mistake or something is going wrong on the job. If the cause is valid, the trade contractor may receive no compensation for the delay. If the cause isn't valid, the trade contractor will claim a change order for delay and an extension of the contract completion date.

The order to suspend the job has to come from the owner. Of course, any owner considering suspension for cause needs counsel from both the CM contractor and an attorney. Suspension for cause will usually be based on failure to comply with the contract, and should begin with a written stop order. The order should cite grounds for a reasonable belief that the contractor isn't following the plans and specs. Identify exactly what part of the plans or specs aren't being followed. Make the order effective after three days so the contractor has time to take corrective action. Deliver a copy of the stop order to the contractor's insurance carrier and bonding company. After three days, the contractor should suspend work until there's an agreement on how to proceed.

> *"If the work is suspended for more than a few days, some loose ends will require attention."*

All of the following can be legitimate grounds to suspend the work:

- recurring failure to correct defective work
- persistent failure to carry out work in compliance with the plans and specs
- chronic failure to prosecute the work in a timely manner and according to schedule
- continual failure to provide adequate supervision of construction crews on-site
- repeated failure to correct excessive disturbance due to site congestion, noise, odors, dust, or interference with other contractors, even if the disturbance doesn't violate the contract
- use of workers with criminal backgrounds or undocumented aliens on public jobs

If the work is suspended for more than a few days, some loose ends will require attention. It's best if the contract covers these issues. Otherwise, each becomes a matter of negotiation — usually at a time when neither the owner nor the trade contractor is eager to compromise. To minimize losses during the suspension, the contractor should:

- relocate materials and equipment to prevent loss, damage and obstruction of passageways
- place temporary enclosures to improve security and weather protection
- provide suitable drainage to avoid water damage
- secure all power and utilities

Suspension for cause assumes the job will be resumed once the cause has been resolved. If the cause can't be resolved, the next step is termination for cause.

Termination for Cause

Both the contractor and the owner have a right to terminate performance of the contract if there's a material breach of contract by the other side. A minor breach of contract is less serious than a material breach. A minor breach gives the right to collect damages but doesn't allow termination.

Some contracts list grounds for terminating the contract. If your contract doesn't cover termination for cause, examine the facts to see if the contractor has committed a material breach. The difference between a material breach of contract and a minor breach of contract is a question of fact. Generally, a material breach is something so important that it defeats the purpose of the contract. Each of the following could be considered a material breach for which termination is appropriate:

- failure to make steady progress, which amounts to abandonment of the work
- refusal to pay subcontractors, suppliers or employees after receiving payment from the owner
- refusing to comply with the contract, law, ordinance or the building code
- bankruptcy or insolvency
- refusal to correct defective work
- deliberately concealing a defect in the work

Termination for cause by the owner ends the right of the contractor to finish the work. After termination for cause, the work will be finished by others at the expense of the contractor. As with suspension of the work, deliver a notice of intent to terminate. Give the contractor five days to either make corrections or deliver written assurance that work will be completed on a reasonable schedule and for the contract price. Deliver a copy of the notice to the contractor's insurance carrier and bonding company. The notice should list the reasons for termination. After five days, deliver a written notice of termination.

After termination for cause:

- The owner takes control of the jobsite and everything on the jobsite.
- The contractor should assign subcontracts, warranties and guarantees to the owner.
- The contractor should remove from the jobsite everything owned by the contractor, including all rental equipment.

After termination, the owner has the right to finish the work at the expense of the terminated contractor. The contractor is entitled to a detailed accounting of the amount due:

> - the unpaid balance due on the contract,
> - less the cost of completing the work,
> - less any liquidated damages provided under the contract,
> - plus the fair value of all materials, supplies and equipment owned by the contractor and used to complete the work.

The contract can make the contractor liable for any deficiency. But the contractor may not be liable if termination was beyond the contractor's control.

Let's Move On

Suspension and termination of contract aren't pleasant subjects. No contractor wants to leave a job unfinished. Fortunately, suspension and termination don't happen very often. The next chapter starts on a more pleasant and useful topic: change orders. You'll see plenty of those on most jobs.

Chapter 11

Assisting with Change Orders

A CHANGE ORDER IS A written modification of the scope of work, contract price, or time for completion under a construction contract. It should include a settlement of all claims for direct, indirect and consequential damages that result from the change. Few jobs go exactly as planned, so changes are virtually inevitable in any complex construction project. Here's why:

- Most plans are only 80 to 90 percent complete. Even if the plans were nearly 100 percent complete, architects and designers are human. When we humans take a second look at something, we almost always find a simpler, better way. In the construction industry, that requires a change order.

- Trade contractors recognize problems others may not see. To get around some problems, extra work will be needed. That's going to require a change order — more money and possibly an extension of the completion date.

- Owners initiate changes. Opportunities for improving the job often become obvious as work progresses. Few owners are expert at reading and understanding plans. They only "get it" when the framing is up. Suddenly, it's obvious that something needs to change.

> Some changes will be the result of a surprise once work starts. For example, field conditions could be other than expected. On a recent job, we discovered the right-of-way for a long-abandoned city street running through the middle of the project.

> The building code or an interpretation of the code could change. Or the building inspector may require some construction detail not included in the plans and specifications. On a school job, I had the state inspector force us to double the size of concrete footings and reinforcing, even though our structural engineer insisted the original design was fine.

> A specified product may no longer be manufactured, or a better choice may be available. In my experience, this is usually discovered two days before you're scheduled to open the facility to the public. Most likely, you need some type of fire alarm gizmo available only from Taiwan. Your choice is to either post a human fire-watch guy 24/7 at $50 an hour for two weeks, or issue a change order and fly the gizmo in from Taiwan. I'll take the change order.

Every CM contractor represents the owner. That makes it your job to negotiate, process and approve changes to the job. This chapter explains how to do that. The next chapter will suggest good ways to protect the owner from unwarranted construction claims of any type.

Since changes are so common, every construction contract should set standards for processing changes. These include:

Prompt notice. It's fair to require that trade contractors give notice immediately when they discover extra work is needed. If the contract requires prompt notice of a claim for extra work, courts and arbitrators usually deny a claim made months after the work is completed.

In writing. Except in a true emergency, every change to a contract should be in writing. That's an owner's best protection against questionable claims for extra work. On a busy construction site, it's easy to forget about getting written approval for changes. What happens if there's no written agreement on a change? Courts in some states hold that an owner who had actual notice of the change and didn't object will be liable for the extra cost. Other states won't enforce an oral agreement for extra work if the contract requires a written change order. Some states require that changes for home improvement work be in writing even if not required by the contract.

The price. Some contracts require mutual consent to the cost of any change. That's most favorable to contractors. Other contracts set guidelines for pricing changes — such as the cost of labor and materials. That tends to reduce the contractor's advantage. We'll have plenty more to say about negotiating the price of change orders.

Is it really a change? Is it something the contractor should have considered when estimating the job? For example, should a trade contractor cover the cost of additional work required by the inspector?

Mutual Agreement vs. Force Account

If you've worked in construction for a while, you understand a contractor's advantage when negotiating the cost of a change order. If the contract doesn't cover the subject, every change will require mutual agreement. The contractor can name any price. There won't be a competitive bid. Take it or leave it. From an owner's perspective, that's an uncomfortable position. Changes by mutual agreement work in the contractor's favor.

Contractors usually insist that work done under a change order will cost the contractor more than work included in the original contract. The administrative cost is higher. More supervision is required. The work schedule may have to be changed. If the owner thinks the cost of a change seems higher than warranted, maybe it's because he simply doesn't understand what's involved.

"If you've worked in construction for a while, you understand a contractor's advantage when negotiating the cost of a change order."

Owners prefer the right to order changes under a set pricing formula. On larger jobs, it's common to handle changes under what's called a *force account*. Changes are the option of the owner. If the owner decides to make a change, the contractor is required to do the work, usually on a time-and-material (cost-plus) basis. Pricing of changes based on cost is more favorable to the owner because the contractor has to prove every item of extra expense.

Other pricing options for force account work include:

> *The contractor's normal selling price.* That can be interpreted as what the contractor has charged for similar work on similar jobs — not easy to either prove or disprove.

> *An agreed cost per unit of work.* This may be a good choice when the only change is additional units of work, such as cubic yards excavated or square feet of concrete slab poured. When the units of work may vary, the contract can set a price per unit more or less than some amount. In that case, no change order will be required if the quantity varies from what was expected.

> *Published prices.* For example, the cost of changes could be based on the price quoted for the most similar item in the most recent edition of the *National Construction Estimator* (or equivalent) plus supervision, taxes, insurance, overhead, and a reasonable profit.

Negotiating Price Changes

Many contractors carry around a perception that all changes have baked-in-the-cake leverage in their favor. As the CM contractor, it's your job to turn that belief into a gross misconception. A long time ago, a salesman I knew told me that negotiating any deal requires the right "snivel." To negotiate change orders, you show that you have some leverage of your own — the snivel. Examples:

"Now I see it all finished, maybe the toilet would be better on the other wall after all. It won't take you a minute to move it, will it? There's only two bolts."

Example 1: "I can settle this now, or a year from now, with lawyers sitting around both sides of the table. It makes no difference to me. I get paid every Friday either way." The CM contractor, not the contractor, controls the project, the schedule and cash flow. You make recommendations to the owner daily. No contractor does that on any project.

Example 2: "You didn't get that check yet! We mailed it six weeks ago." Nothing gets compliance faster than a little economic drag. Any 60-second decision can be stretched out over two weeks. Rejection can take another two weeks; right after the owner gets back from vacation.

Example 3: "The owner's wife's sister's aunt doesn't like the 14th carpet color sample you submitted. Could we have another 30 samples, please? And, oh, while you're at it, the owner tells me he left that big box of submittals you gave him in his rental car at the airport and they've gone missing. He needs six more sets as soon as you can get to it." Paper processing fatigue will clear the sinuses of any contractor who thinks he has change order leverage.

Example 4: Recently I was an expert witness for the defense in a million-dollar claim by a contractor who refused to submit backup data for labor and material charges. The case wandered around in the halls of the courthouse for almost two years without going to trial. Meanwhile, the great recession cut off the contractor's source of income. The lawyers swooped in with all their requests for discovery. While the contractor's cash flow was drying up, his legal carrying costs soared. It was a simple case of who had the deepest pockets. The defense didn't need to take the case to trial. To win, they only had to grind the contractor up against his legal carrying costs. When he had drowned financially, settlement was easy.

The snivel leverage list is endless. Of course, *I* wouldn't resort to such tactics. But I've heard of people who have.

There's always going to be wiggle room in the contractor's first offer. As a CM contractor, it's your job to figure how much slop is floating around inside the price quoted for any change order. Finding the slop money requires two things; a good relationship with the contractor and a little polite and tender probing.

Assuming you've got the relationship covered, try offering 50 percent less than the asking price. If that doesn't fly, offer to buy lunch. If the contractor insists on paying for lunch, you've probably offered less than the actual cost of the change. "OK, tell me what you can do cost-wise on this. I'm sure not taking your first offer to the owner." Usually the contractor will come down 5 percent. I ask for 25 percent. We end up at 15 percent off the first offer. Do that dance 30 times on a project and you've bought the owner a new car and yourself your next job.

Time-and-Material Cost

Pricing extra work on a time-and-material basis can be complex. Anything omitted from the calculation gives the owner a discount. So contractors have to account item-by-item for every cost, including overhead. If the cost is too high, you've got alternatives:

> ➤ *Materials* — Offer to have the owner supply what's needed.
> ➤ *Labor* — Look for a clause in the contract which allows the owner to do the work.
> ➤ *Subcontract* — Look for a clause in the contract which allows the owner to hire separate contractors.
> ➤ *Equipment* — Have the owner supply rented equipment.
> ➤ *Overhead (both jobsite and off-site)* — This is a percentage and should be set in the contract.
> ➤ *Markup* — Again, this should be set in the contract.

Changes can either reduce or increase the amount of work. Some changes can result in a credit rather than a charge.

Material cost should be limited to the actual cost to the contractor. That includes transportation, storage, sales tax, drawings, warranties, bonds, fuel, temporary construction, scaffolding, utilities and delivery. But if there's any trade discount, be sure that's deducted from the charge. Estimate the cost of consumable supplies such as tape and rags. Reducing the materials required may not yield a credit equal to the original cost per unit. Vendors often add a charge for restocking or partial cancellation.

Labor cost should include taxes and insurance based on payroll and employee benefits for everyone required to complete the change order, including jobsite supervision. Whether required by law or a collective bargaining agreement (the *prevailing wage*), the following are considered labor costs:

> health care
> compensation insurance
> liability insurance
> payroll taxes (state, federal and local)
> union, apprentice and pension costs

For a major change order, ask for a detailed breakdown of the cost of each labor classification. Hourly labor rates should be the same as the contractor pays tradesmen for similar work on the same job. Question the use of any special labor classification for change order work.

Subcontract cost should be based on a signed copy of the subcontractor's detailed invoice showing the charge for labor, material, equipment and markup. The subcontractor's overhead, profit, taxes, indirect supervision, insurance, bonds and warranty shouldn't exceed 15 percent of the direct cost (labor, material and equipment).

Equipment cost should be limited to the actual cost of equipment required. Exclude any cost incurred when equipment isn't in use. Equipment expense shouldn't include the cost of moving equipment to the site if the equipment is required on-site for some other purpose. Exclude the rental cost of tools and equipment valued at less than $200. Estimate the cost of consumable tools, such as brushes, rollers and drill bits. The charge for equipment owned by the contractor should be competitive with rates charged for similar equipment at rental yards in the area.

Overhead expense can include estimating and purchasing, indirect supervision and project management, home office overhead,

---

Changes in Materials

Not every change requires a change order. Even changes at extra cost to the contractor may not require a change order. Here's an example.

A few years ago, I was the CM on a high school gymnasium project that required a special kind of vinyl floor tile. The contractor said there wasn't enough of that tile in the supplier's San Francisco warehouse to do the whole job. But they could get the rest from other warehouses in Chicago or Atlanta.

When you need an acre of floor tile, it's not a good idea to order from different warehouses. Most likely, you'll have different dye lots. It's not going to match.

The contractor reassured me: "We're running short on time. This may be special tile, but it's just your basic gray. It'll match."

I said, "OK. It might match. So I'll make you a deal. If you want to place multiple orders from Chicago, Atlanta and San Francisco, go ahead. But if the owner and I can tell the difference and don't like what we see, you'll have to pull it up and reorder. As the owner's rep, I'm only going to pay for that tile once."

Well, they ordered from three warehouses. The tile came in. They put it down and we had a look. It didn't match! The contractor had to air-freight an entirely new, single-dye lot batch of floor tile. That tile then looked great. We finished the job just 36 hours before the big open house.

The tile required extra work and extra freight cost. But there was no extra cost to the owner.

Cardinal Changes

I know a contractor who took on a small job on a ski chalet at a popular winter destination in Colorado a few years ago. The owner wanted a new sundeck and some interior work in an unfinished basement. It would be a few weeks work. My contractor friend wrote up a bid, offered his usual agreement, got the owner's signature and started work.

By mid-October, the deck was done. The owner was pleased. My friend started on the basement — another week of work at the most. But first, the owner wanted some changes — $5/8$-inch wallboard with vinyl finish. My friend wrote up a change order and got a signature — at the cost of labor and materials plus 10 percent, exactly as required by their contract. Before the wallboard was up, the owner added Pergo flooring to the job. No problem. My friend wrote out another change order and got a signature, again at cost plus 10 percent. Before the floor was done, the owner wanted another change. This time he wanted a bar built into a corner of the basement. My friend wrote out another change order at cost plus 10 percent. Before that was done, the owner decided that the family room upstairs needed Pergo just like the basement. They signed another change order at cost plus 10 percent. By this time it was past Thanksgiving.

You can imagine the rest of the story. My friend worked in that chalet nearly all winter – at cost plus 10 percent. Finally he asked me if there wasn't some way out of the cost-plus-10-percent contract. He was working for wages at a time when good-paying contract work was plentiful.

He should have asked sooner.

The law calls it the *cardinal change* doctrine. Changes to a contract have to be within the general scope of the agreement and have to be relatively small changes. Large changes (or too many small changes) are considered a *cardinal* change and have to be the subject of a new contract. My friend got into cardinal change territory somewhere between the wallboard and the Pergo.

change order negotiation and processing, course of construction insurance, clerical, and purchasing expense. A change which yields a net credit to the owner is unlikely to reduce overhead expense. So don't deduct anything from overhead if the change results in a net credit to the owner.

Markup on direct costs should be limited to 25 percent or less, depending on the size of the project. Direct costs include all work done by a contractor's crews. Markup on subcontract work shouldn't exceed 15 percent. Don't make any deduction for markup if the change results in a net credit to the owner.

Other Change Order Issues

Negotiated change orders can be good business for contractors. Contractors have been known to take work at cost on the expectation that negotiated changes will provide a comfortable margin. Watch out for the following:

Price changes. On a job that lasts more than a few months, you can expect that prices for both labor and materials will change before work is complete. Unless the contract provides otherwise, price changes alone aren't considered grounds for a change order. If the job requires more than about six months to complete, the contract should provide that price changes alone aren't grounds for a change order.

Extending the contract time. Extension of the completion date is considered a benefit to the contractor and shouldn't be the basis for a claim for extra compensation. Any request for extension of the contract time should include a detailed schedule showing milestones for the extra work. Don't approve any time extension that falls within a previous delay claim period. You could end up paying for time that was already allowed for in the delay claim.

Weather delay. Contract completion dates can be assumed to include usual weather delay. Bad weather that exceeds what can be expected may be a cause to grant additional time to complete the work. But weather delays should not inflate an owner's cost.

Define weather delay early in the project. Don't leave it up to nature to decide. If your jobs are in the Sun Belt, a week or 10 days in the schedule should cover weather delays on most jobs. Anything beyond that is extra weather delay. There shouldn't be any reason for discussion. Put weather delay days, both used and remaining, at the top of your job meeting minutes. Deduct weather days as they happen. That leaves no doubt about the status of weather delay.

"You know, if it weren't for change orders, I couldn't send my kids to college."

Failure to keep records. A contractor doing changes on a time-and-material basis has to keep separate cost records for extra work: extra time required and the extra materials used. Estimates aren't enough. You need to see receipts and time sheets. If the contractor can't show a record of extra costs, your contract should allow payment based on a good faith estimate by the *owner*. The same applies to extensions of the contract completion date. In the absence of good records, the new completion date is set by a good faith estimate of the owner.

If I can't cut a deal on the cost of a change I disagree with, I crawl through every nut, bolt, nail and timesheet. Don't assume anything. Insist on seeing the original records. If the contractor can't produce those records, I'm not going to approve the charge. About halfway through this analysis, the contractor usually wants to cut a deal.

Subcontracts. Subcontracts should also require subcontractors to proceed with changes required of the prime contractor. Recordkeeping requirements for prime contractors should also apply to subcontractors. For example: "Any claim by a subcontractor for extra work shall be made with the prime contractor and include receipts and time records supporting the claim. Only the prime contractor has the right to make a claim with the owner for changes in the work." The prime contractor has the signed contract with the owner. Subcontractors don't have a contract with the owner and have no right to make claims against the owner.

Notice. To avoid surprises, insist on a written proposal before any extra work is authorized. A contractor who feels that some instruction is a request for extra work should submit a proposal describing the change and the cost. Authorization is required before beginning work.

On fast-moving projects, request a *ROM* (rough order of magnitude) estimate when you get notice of a change. Get a ballpark estimate of what the work will cost *all in* — with overhead, profit, everything included. If you can stomach that cost, put it in an email with instructions to proceed on a time-and-material basis against that ROM number. To get paid, the contractor has to document all labor and material costs — receipts, time cards, etc.

Minor changes. Both the owner and the contractor can make minor changes which do not (1) materially alter the quality of work, (2) don't affect the cost or time of performance and (3) comply with applicable laws, codes, ordinances and regulations. Substitution of one product for another will be OK, assuming both have similar characteristics, value and cost, and assuming the substitution has been approved as required in the specs.

Payment for changes. It always simplifies accounting if payment for changes is due in full when the change is complete. If the change is done over several pay periods, charges will be billed over those pay periods as work proceeds. A contractor who accepts final payment for extra work waives any claim for additional pay on the work completed.

Failure to agree on changes. If you can't agree on a price for a change, settle the issue as a construction dispute. That's the subject of Chapter 12. Failure to agree on changes shouldn't delay progress of other work. As a show of good faith, consider paying some agreed-upon percentage of the amount requested. Sixty percent may be about right. Agree to negotiate the balance at the end of the project. That way, the contractor recovers most of the cost of the change. The owner appears to be fair. And the work doesn't slow down.

Allowances. If the actual cost of some item in the job isn't known when the contract is signed, the contract may include an approximate allowance. For example, the contract might allow $5,000 for light fixtures, to be selected later. If the actual cost is more or less than the allowance, adjust the contract price with a change order.

Processing Change Orders

Requests for change orders should be routed through the CM contractor. If you allow the owner to negotiate changes directly with the contractor, you've lost control of both the job and the owner. The *Change Order Agreement* in Figure 11-1 should work for most changes on a small job. Use this form if you don't have a form that works better.

Change orders can originate with any contractor or subcontractor, the owner, or a designer. If the contractor is making the request for a change, insist on a cost estimate and the likely effect on the completion date. That's true whether work will be done at a negotiated price (requiring mutual agreement) or under a force account (such as cost-plus). Will written plans or changes in the specs be required? If the contractor has questions or objections, address those issues before going further.

Change Order Agreement

Today's date _____   Original contract date _____

Job address _____   Original contract price $ _____

City, ST, ZIP _____   Sum of previous changes $ _____

Job phone _____   Cost of this change $ _____

Contractor _____   Revised contract price $ _____

Description of this change: _____

_____

A. Material and supplies cost: $ _____

B. Taxes and fees: $ _____

C. Direct labor: $ _____

D. Indirect labor costs: $ _____

E. Equipment and tools: $ _____   **F.** Subtotal (A to E): $ _____

G. Overhead at ____ % of line F: $ _____

H. Subcontracts: $ _____

I. Overhead at ____ % of line H: $ _____   **J.** Subtotal (G to I): $ _____

K. Profit at ____ % of lines F & J: $ _____

L. Total cost, lines F, J & K: ☐ Add   ☐ Deduct $ _____

Items specifically excluded from this change: _____

_____

☐ This proposal is valid for ____ days.   ☐ This change deletes for credit:

☐ We require ____ days extension of the contract time.   _____

☐ We are proceeding with this work per your authorization.   ☐ Deduct for retainage: _____

☐ Please return a signed copy of this agreement as your acknowledgment of this change.

This Change Order incorporates by reference the terms and conditions of the original contract and all change orders approved prior to the acceptance of this agreement.

This Change Order is accepted by _____   Date _____

Figure 11-1
Change Order Agreement

> **A Shortcut for Change Orders**
>
> On larger jobs, architects often write change orders. That's seldom a good choice, as change orders are low priority work in an architectural office. The architect will have to get cost information from the contractor, which can delay approval of a change for weeks. There's a better way.
>
> I hate paperwork. So on jobs where I'm the architect, I cut a deal with the contractor. The contractor only wants to build and get paid, and I don't want do any paperwork I can avoid. Since the contractor already has change order cost information in his computer, he probably has a change order form handy. That's everything needed to prepare a change order, except one thing, motivation. Here's how I supply that motivation: "You write change orders for my review. Do it the day you know a change is needed. I'll have a decision back to you in hours, not weeks. So it's your call, my way or the old-fashioned way — I've got nothing but time."
>
> It's an easy choice for a contractor. Speeding up the change process keeps the money flowing. I don't make any promise about accepting the contractor's prices. This isn't a bribe; it's project expediting. It's an exchange of paperwork for progress, not an exchange of money for progress.

Don't end the discussion about a change without something in writing. Define the work, the money, the time and the responsibility. Then get a signature from somebody authorized to commit to the deal. Even if it's only on a napkin or sheet of copy paper, get the basics written, signed and dated. It won't be the first or last contract written on a napkin. Then give everybody, including the owner, a copy. When you have more time, transcribe your notes onto a formal change order.

Circulate that proposed change order to each contractor involved, and to the designer or engineer if there's a change in design. When you have a consensus on the proposed order, send a copy to the owner with your recommendation for approval, rejection, or modification. The owner signs the change order only after everyone else has signed. Your recommendation to the owner should answer these questions:

- Is the change really necessary?
- Is the quoted price reasonable?
- Is any change in the schedule reasonable?
- What impact will the change have on other work?
- Will there be any additional costs from making the change?
- Will plan review and an amended building permit be required?
- Are there alternatives that serve nearly as well, at a lower cost?
- When will payment for the change be due?

Assuming the change is approved, help prepare any documentation required to implement the change. Then relay written instructions to the contractor to begin work on the change.

When signed by the contractor and owner, a change order becomes an amendment to the contract and should be filed with the contract. Be sure all contractors concerned, the designer, and the owner receive copies of the change, including additions to both the plans and specs. If the change requires plans or specs, the architect should also sign the change order.

Always keep a log of change order requests, even if it's just one page in your log book: include the date requested, a summary of the change, your recommendation, the date either approved or declined by the owner, and the contract change both in dollars and time.

A Final Appeal on Changes

Don't let change order claims go beyond 30 days. Settle claims as they show up, not as they pile up. Let claims linger for six months after they occur and you deserve whatever happens. I try to settle a claim on the spot, the moment it comes up. If that's not possible, settle the claim no later than the next job meeting.

Require the contractor to submit claim paperwork quickly. If they delay, there's a good chance they'll forget something, and by the time they find it, construction will have moved beyond that claim. Most contractors won't make a second claim against a first claim, so it's in their own best interest to be prompt. If they choose not to, and forget something, let it be. Let that happen enough times on a job and you've saved the owner some real money.

Things happen fast on a construction site when the players have the authority and the motivation to make decisions. It's fair to ask a contractor to make a commitment on the spot if you're ready to do the same. With the owner's full faith and support, there won't be any delay on your side of the fence. When you understand the problem, draw a solution on a yellow pad, a piece of drywall, or on the plans. Or spray paint it on the ground. Then ask, "We're ready to write a check. Why aren't you done yet?"

That's how projects get finished early and on budget. That's also how you spell HERO to an owner.

Value Engineering

Wise property owners encourage contractors to identify opportunities for value engineering. Contractors can spot ways to reduce costs or improve results that aren't obvious to an owner or a design professional. Many construction contracts offer a financial incentive for value engineering proposals that become part of the job. The usual incentive is a reduction in the contract price equivalent to 50 percent of the savings. For example, suppose a contractor offers a value engineering proposal that will save $10,000. If that proposal is accepted, the contract price is reduced by $5,000. The result is a 50-50 split of the savings.

Every change order and every proposal that could result in a bonus has the potential to end in a dispute. That's also true of change orders resulting from a value engineering proposal. For example, it's easy to consider a value engineering proposal to be obvious once made. "Why didn't I think of that?" When that happens, it's hard to award a bonus amounting to thousands of dollars.

To avoid disputes, be sure your contract includes very specific rules for making value engineering proposals. For example, any value engineering proposal a contractor submits should include:

- a full description of the proposed change
- a comparison of the advantages and disadvantages in service life, reliability, economy of operation, ease of maintenance, design safety standards, appearance, and impact on the contract completion date
- a list of changes required to the contract, plans and specifications if the proposal is adopted
- an offer to complete the proposed change at a specified price
- the time when a decision on the proposal is required to get the forecast benefit
- if required, a pledge that proposed changes will be prepared by a licensed professional architect or engineer

If the proposal is approved, the contractor will have to:

- sign a consent to the change, the new design, design calculations and design criteria
- submit a detailed cost breakdown for the proposal, including cost to subcontractors and the owner for architectural, engineering, or other consultant services, and the staff time required to examine and review the proposal

All costs of developing and implementing the proposal have to be paid by the proposing contractor. The owner gets the other half of the savings.

Details on Value Engineering

Since value engineering proposals can result in a dispute, avoid the most obvious risks by covering the following issues in your contract.

Protection of the idea. The contract should offer some protection against the owner using a proposal or some variation of the proposal without paying compensation to the contractor. For example, "No part of this value engineering proposal shall be disclosed, duplicated or used for any purpose other than to evaluate this proposal."

Cost of the proposal. The contractor should cover the cost of preparing a proposal, including design. The contractor doesn't collect until the work is done.

Nothing happens until the proposal is accepted. The contractor is obligated to do all work in the original contract until the value engineering change order is adopted.

> **Value engineering**
>
> Finding ways to reduce costs, speed completion or improve the result without loss of intended utility and without reduction of desirable characteristics such as service life, reliability, economy of operation and ease of maintenance.

Delay. Processing of a value engineering proposal isn't an excuse for delay.

Rejection. The decision to approve or reject any value engineering proposal should be made at the sole discretion of the owner. For example, a proposal can be rejected if it includes disadvantages. The contractor has no claim for reimbursement of costs if a proposal is rejected.

Value Engineering in the Design Phase

To this point, we've been discussing value engineering after the contract has been awarded. But that's not the only opportunity for you to minimize cost and maximize value. In my opinion, the best value engineering is done when the plans are being drawn.

Convince the owner to hire a contractor on an hourly basis as a consultant when the architect is hired. Put the architect in charge of the owner's design. Put the contractor in charge of costs. When the architect has a preliminary design ready, get the contractor to value engineer the price — by the next meeting. The cost estimate the contractor brings to the next meeting should be in the form of a bid ready to be accepted. That keeps all eyes focused on savings and the contractor's price pencil sharp.

Let a contractor do the value engineering for you. The price on bid day shouldn't be a surprise to anyone. More likely than not, the owner will end up with a plan that can be built on schedule and within budget on the first bidding, not the third.

I realize that architects routinely draw up estimates. But I've never seen an architect offer to finish a job for the estimated price. Ask an architect when he last bought a toilet, and how much it cost. Ask the contractor the same question. Compare their prices, and then decide who gets paid to do your cost estimates.

Where We Stand

Before moving on, I want to make a pitch for competence. Every CM contractor has an obligation to be the best-informed professional on the site. Know the project better than the designer and the contractor combined. The fact that you're reading these words is good evidence of your dedication. Understand the plans. Live and breathe the construction process. Don't take on any project unless you're willing to make a 100 percent effort.

Having made that point, let's move on. This chapter and the next cover related subjects. Much of this chapter assumed the contractor had made a claim for extra work. The next chapter describes what a CM contractor should do to protect the owner from claims, including claims for extra work.

Chapter 12

Protecting Against Construction Claims

EVERY CONSTRUCTION PROJECT COMES with risk of loss, usually a loss to the owner. Many projects run well over budget. Some sit idle or incomplete for months or even years. A few are never completed — just abandoned or demolished rather than finished. Plenty can go wrong on a construction site.

The person with the most skin in the game is nearly always the property owner. Of course, the lender and the insurance carrier have an interest as well. But the essence of their business is taking calculated risks. No property owner starts a job weighing the chance of a financial disaster. But wise property owners understand that surprises are common on construction projects. The best way to manage that risk is to rely on the skill of experienced professionals. That's why risk management is part of every CM contractor's portfolio.

Reducing the risk of claims starts with design. The more conventional the design, the more likely that design can be executed within a reasonable period and for a predictable price. The more complete the plans and planning for the job, the less likely an expensive surprise.

Architects are notorious for running up project costs with their fancy designs and elaborate extras. I can admit that because I'm a licensed architect and in my earlier years I've done that myself. But in today's green-design era, *waste not, want not* is the more appropriate way to go. As a CM contractor, it's your responsibility to coach the project architect. The design shouldn't consume all the framing material in three lumber yards and the total production of a small concrete plant. To reduce the owner's risk of loss, design around limited resources and limited money. Conserve both materials and labor and you'll save money. For me, the first rule of LEED (green) construction is the conservation of labor and materials. For example, why use a hip roof if a gable roof will serve the purpose just as well? And, you don't need to make a room 12 feet high if a 10-foot-high room will give you the same effect. Keep in mind that designing with less will ultimately get you more LEED mileage moneywise than buying LEED materials.

Adequate insurance coverage, good records, jobsite safety, quality assurance (QA) and an effective policy on change orders also reduce the risk of loss. We covered insurance, bonding, safety and QA in Chapter 7. Chapter 8 identified the records you need to collect and preserve. And, we covered change orders in the last chapter. Taken together, these subjects don't cover all the concerns of risk management. Construction claims arise out of contexts entirely separate from these. This chapter will cover the remainder of the subject of risk management. We'll also suggest how to handle claims that can't be resolved by negotiation.

Allocating Risk of Loss

We're not going to suggest that all risk of loss can be shifted from the owner to designers, contractors and subcontractors. That isn't possible and wouldn't be good policy even if it *were* possible. A better approach is to consider who should carry the risk of loss at the most reasonable cost. When that isn't the owner, distributing the risk of loss to others is good policy for the CM contractor and the owner. As we said earlier, the owner has enough skin in the game already. There's no advantage to piling that risk on any deeper.

As a matter of public policy, risk should be allocated to the party best able to control the loss. That simply makes good sense. It's a matter of taking responsibility for your own actions. Here's an example.

> There was a time when owners and contractors routinely used indemnity (reimbursement) clauses in contracts to make subcontractors liable for any loss to the owner or contractor — regardless of fault. Subcontractors were at the bottom of the pecking order. They had no choice. Either sign a contract accepting liability for any loss on the job, regardless of who was at fault, or look elsewhere for work. That makes no sense if the owner's negligence or the contractor's negligence is the cause of the loss. In fact, indemnity clauses may even encourage recklessness.

Many states now void indemnity clauses in contracts if someone would be indemnified for their own negligence. A contractor or owner who can reduce the risk of loss should be required to absorb any loss that occurs.

Much of this chapter is about claims, including how to protect the owner from financial liability. Before we get into that, we need to cover a CM contractor's professional responsibility when a claim happens. What should you do when you receive, or expect to receive, notice of a claim that's unlikely to be resolved on the spot?

1. begin collecting information on the claim
2. notify the owner about the claim
3. do what you can to mitigate the loss
4. begin planning settlement

We'll cover each of these in more detail.

Collect Information on the Claim

Memories fade — sometimes into outright amnesia. Records become hard to locate with time. Conditions change. The best time to collect information on a claim is the moment you receive information that a claim may be pending. Take pictures, talk to those involved, ask questions, get statements, collect names and contact numbers.

We've emphasized the importance of recordkeeping, and suggested keeping your own personal legal file and a bound job log. These recommendations go double when a claim is developing. Personally, I try to avoid paperwork and filing if I can. But I always get aggressive about collecting information if there's the slightest hint that a situation could turn into a court case down the road. I document the weather, time, place, who, what, where, when and how. I take dozens of photos. I make a very complete verbal recording of what's going on, to be transcribed later if necessary. I gather plans, specs, handwritten notes on drawings, and emails. I do short interviews with all the players involved and transcribe a summary to my bound ledger.

When a legal fight breaks out, you need to be the one with the most complete and best-organized information. Any time the issue is legal, collect your information *right now, today*! Be the leper with the most fingers. Contractors may lose critical files. You don't want to lose anything!

Notify the Owner

Notify the owner within the hour if you discover a potential claim. Bad news travels fast. Never let the owner hear news from someone else first, especially bad news. If you're not the first to share bad news, your job could be yesterday's news. Withholding information is always a mistake, even if you're in agony about delivering that news. Get it over with today! Then get a good night's rest, and plan to write the report in the morning.

I like to deliver good news along with the bad. The good news is that I have a plan to backcharge the potential claimant. I call this the *self-canceling claims defense*. Here's how it works:

When there's a claim, I meticulously crawl through the past performance of the company making the claim. Look for places where they were late in providing information, materials or labor. Time is money. I convert lost time into a value I could charge to the claimant. Often, something that's turning legal is also something that's turning into leverage. Many times leverage will triumph over legal, given sufficient time.

When plans don't arrive on time, the job is delayed. When lumber arrives a week late or a subcontractor doesn't show for 10 days, that's a delay. When a contractor delays ordering long-lead-time items, there's another delay. The owner's time has a value just like the contractor's time has value. We all have only so many years to live. What hourly rate would you place on that time?

The owner's first question will be about who's involved, and the total exposure. Answer those questions. Then suggest that our side has a countervailing claim. You're collecting information on that claim. Details will follow, both direct costs (time, equipment, material) and indirect cost (home office overhead).

Mitigate the Loss

For example, if the claim relates to contractor delay, do what you can to get work under way once again. When the claim has matured (the value of the claim can be determined), give the owner a full report, including a dollar estimate.

When projects bog down, people get excited about making claims. Understand that most claims are like Jell-O, soft, squishy and amorphous. A CM contractor can do a lot to defuse the situation. I recommend gathering information, finding the leverage, and keeping your composure:

➤ *Information.* When you know more about the claim than the claimant, you'll be more in control of the outcome.

➤ *Leverage.* The claimant needs money. You're in control of the cash. That's your primary leverage.

➤ *Composure.* Sit down with the claimant. Offer a sympathetic ear. Listen calmly as the claimant vents. There's nothing like knowing the case from the other side when planning your defense.

Here's an example of using information, composure, and leverage to carry the day:

A few years ago I was the CM contractor on a school board project — a track and football field estimated at over $1,000,000. It was dragging a bit, timewise. The prime contractor had screwed up here and there when installing irrigation and drainage. Things were getting a little tense. The prime contractor demanded change orders for extra work and an extension of the contract time.

We'd run through most of the school's budget. I had just enough left to complete the project — assuming nothing went wrong. But we still had 90 days of work ahead of us. Writing up the change orders the prime contractor was demanding would soak up our remaining contingency, leaving us with no margin for error. I wasn't going to let that happen. So I called a meeting of everyone concerned, including representatives from the school's bond management team — over 20 of us in all.

Before the meeting, I plotted a strategy with the representative from the bond-management team. I asked him to be a spectator at the meeting. I always tell the owner that only one of us can negotiate a deal. If we both try, the claimant will divide and conquer us, so his sole responsibility at the meeting was to nod approvingly any time I made a point.

I started the meeting by calling on the prime contractor to air his claim. I let all of his subcontractors do their whining as well. When everyone had their say, I did some explaining.

"What you guys say about your costs is absolutely correct. I've got no argument." The owner nodded his approval.

"On the other hand," I continued, "you're also late with the work. You've screwed up some. What you're asking for, I'll never be able to get approved. The board will never let me continue this project with no contingency." I paused to let that sink in. Most in the room turned to look at the owner nodding his approval.

"We'll have to shut this job down and start talking to lawyers." By now, everyone was watching the owner. On cue, he nodded his approval.

"So, here's what I suggest. I have to hold $100,000 in reserve. But I'm willing to release the rest now, if that's what it takes to finish this job. Take that proposal back to your ant hill. See if your boss would rather finish this job in the next 90 days or litigate for the next five years." You could just smell the leverage floating in the air. I glanced at the owner to be sure he was nodding his head. He was.

The next day I got a call from the prime contractor. Everyone had accepted my deal. I didn't really have to take my deal to the school board. The bond-management team representative was sitting in the meeting. The project finished on time and within the budget. The contractors were semi-happy with the outcome; although they had to swallow their claims, we gave them a super recommendation on other work later.

Begin Planning Settlement

Plan your settlement right from the start. Read the contract for settlement procedures. Decide, with advice from the owner, on the best way to resolve the claim. No project is complete until all claims have been resolved.

Settling claims can take a year or more. Settlement over such a long period is beyond the resources of most contractors. The cost of legal counsel can run $300 an hour or more, plus expenses. Divide a claim for $100,000 by even that rate and you get 333 hours. Then consider expenses and you're down to perhaps 250 hours, a little more than six weeks of an attorney's time. Settling any significant dispute will require that much time or more. The claimant will always be the first to begin paying legal bills.

It's just plain dumb to spend $100,000 in attorneys fees chasing a $100,000 claim. It makes even less sense to pursue smaller amounts. For a claim to be worthwhile, your suit needs to approach a million bucks.

I don't recommend making claims against an architect. That's not just because I *am* one. Most architects are just a project or two away from unemployment. Many don't carry Errors & Omissions Insurance. All they have to lose in court is their house, car, and a couple thousand dollars in savings. Nearly all of that can be protected by filing bankruptcy. Most architects are judgment-proof that way. You may win the claim, but you won't be able to collect a dime. That's wonderful bug repellent when annoyed by litigation. Get a judgment if you want. It won't do any good. Like any other debt, judgments expire eventually.

An architect with Errors & Omissions Insurance will pay in his deductible and watch from the sidelines while an insurance carrier's attorney defends the claim. You'll be left battling a giant insurance company, while the architect is a spectator in the legal process.

Primary Source of Claims: Surprises

Contractor claims can be based on deliberate acts, such as interference by an owner, architect or engineer. Claims by subcontractors and vendors usually relate to slow payment or nonpayment. But the most serious claims will be based on some type of surprise — something that wasn't anticipated by the contractor or a subcontractor. It could be a mistake in the plans or specs, something discovered on the site, or something required by the building inspector. We'll cover all of these. But all have one thing in common: they're unanticipated — not part of the contractor's bid.

Once work starts, nearly every surprise will add to the cost of construction. Very few surprises are welcome news. Every significant surprise can result in a claim against the owner. In this chapter we'll explain what a CM contractor can do to minimize an owner's risk of construction claims. In this context, remember what I said about all plans being incomplete. Even the most complete plans omit about 20 percent of what's required to finish the job. That 20 percent includes plenty of surprises that a CM must try to keep from turning into money.

First, understand the importance of the contract in this context. Construction contracts and subcontracts are written to allocate risk — divide the benefits and burdens between the owner and the contractor. Courts in all states recognize allocation of risk as a fundamental purpose of construction contracts. Courts very seldom re-allocate risk after a loss occurs. All courts enforce contract terms as written unless prohibited by statute or overridden by some state interest (public policy). What your contract says about surprises and construction claims will almost certainly be enforced by the courts in your state.

So take your time with the contract. Good contracts anticipate the most likely disputes. The best contracts resolve those disputes in your favor.

Anything you can do to reduce a contractor's opportunity to claim a surprise will help protect the owner. A claim for extra work is weak if there's no surprise. A contractor, subcontractor or vendor fully and accurately informed of job conditions before the contract was signed isn't going to make claims. To protect the owner, be sure prospective contractors on the job:

➢ are fully informed about site conditions,

➢ have ample opportunity to study the plans,

➢ get prompt and accurate answers to questions about the job.

You can't do anything to inform subcontractors of their responsibilities. You have very little control over relations between the prime contractor and subcontractors. But you can insist that the contractor relay instructions to subcontractors promptly. How information is passed to subcontractors and how work is divided among subcontractors is strictly a matter for the prime contractor. Subcontractors don't have the right to make claims directly against an owner. Subcontractors can make claims, but only through the prime contractor. A subcontractor's primary leverage on any job is the threat to file a stop notice against the prime contractor.

Site Walk

Claims for extra work are weak when site conditions were known or should have been discovered with reasonable investigation. It's simply poor estimating practice to bid a job without evaluating the site conditions. Whether or not a contractor visits the site, it's appropriate to charge a contractor with full knowledge of everything that's obvious at the site. For example, proximity to neighbors, limits on site access, poor drainage in wet weather, overhead power lines, limited storage space, a high crime rate, lack of parking, absence of local retail outlets, the distance to a power source, the economic and regulatory environment, and dozens of other conditions should be apparent to any alert and conscientious contractor. Any claim of extra work based on obvious conditions has little merit.

I generally insist on a mandatory pre-bid site walk. When I conduct the walk, I pass around a sign-in sheet with spaces for company names, addresses, telephone numbers and email addresses, date of site walk and site location. I disqualify from bidding any company that didn't attend the job walk. That may limit the number of bidders, but those left will be serious bidders. None will be able to say later, "Gee, I didn't know that was there. We didn't allow money for that."

More Surprises

On jobs that require extensive excavation, the owner will usually provide the results of soil borings and tests when distributing the bid package. Bear in mind that soil tests and borings establish the general makeup of the soil and bearing capacity for foundation design purposes. A soils report doesn't mean anybody has the slightest clue about what else is under all that dirt. All sorts of tricky things can happen with soil. You may know there's bedrock 10 feet down where they did the boring and soils test trenches, but 20 feet away, it could be sandy clay with the bearing capacity of a roller coaster in an earthquake.

Or, there could be dead Indians under there. I've found them. When that happened, I had three choices:

> ➢ abandon the project
> ➢ move the building 100 feet west
> ➢ bring in an archeologist to excavate the site under the watchful eye of the tribe's medicine man

Millions of miles of utility lines have been buried all over the U.S., starting from the time of Abe Lincoln. Many of these lines aren't shown on any map. Even a utility survey expert isn't going to find all the lines on some sites. Lines that aren't metal don't show up well on ground-reading radar. I can't count how many times I've dug up a water line, sewer line or old septic tank on a construction site.

Today, we routinely lay a metal wire on top of nonmetallic underground piping. That makes future tracking with radar easy. But with the old stuff, it's always a crap shoot — and will probably end in an extra's claim in your next month's billing statement.

Voluntary Disclosures

The site plan should identify utility access points and utility easements: water, power, gas, phone, cable TV, waste, drain or gas lines. The law in some states makes the property owner liable for damage done by a contractor who isn't informed of utility lines known by the owner to exist on-site. Courts usually find the plans to be defective if the owner has concealed an error in the plans or specs that the contractor was unlikely to discover. Don't take that chance.

Reveal in writing everything you know about the site that could delay work or add to the contractor's cost: mold, structural pests, hazmat, subsidence, zoning restrictions or setbacks, requirements of any property owners' association, environmental protection regulations, unusual building code requirements, any prior site survey, previous attempts at similar work or known obstructions. If as-built plans are available from prior construction, make those plans available to the contractors before the bid opening. You don't have to research all these issues yourself. But don't leave yourself open to a claim of deliberately withholding relevant information.

You'll need the owner's cooperation to make full voluntary disclosure. Review the list of disclosure topics in the previous paragraph with the owner to help jog the owner's memory.

Errors in the Plans

Errors are common in plans and specs. Some apparent errors can be resolved by rules of interpretation, as explained in Chapter 10. Conflicts, ambiguities and omissions that can't be resolved by rules of interpretation should be resolved as early as possible with a clarification from the architect or engineer.

"Oh, oh. I think the architect has this picture window on the south side."

Contractors aren't licensed design professionals and don't have an obligation to find errors or omissions in plans and specifications. But all contractors have an obligation to request a clarification when the plans or specs aren't clear. The best time to request a clarification about an obvious error is before the contract is signed. If an error isn't obvious, a contractor is entitled to rely on the plans or specs. If an error or deviation from the building code is obvious, the contractor has an obligation to notify the owner's representative. If the plans are changed as a result of that notice, a change order (including an increase in the contract price) will usually be required.

The right to rely on the plans is no defense when a contractor continues work after discovering an error. Contractors are liable for avoidable costs and direct damages if they continue work after recognizing the need for a clarification. Failure to request a clarification in time to avoid delay in the work can make a contractor liable for both lost time and wasted materials. When you need to make a change in a hurry, consider using the phone and fax procedure I recommended in Chapter 7. It's the best way I know to get design information back and forth between designer and contractor.

Notice of Discrepancy

All contractors compare what the plans show with what actually exists on the jobsite. Often, a contractor will find a discrepancy. Some state courts enforce contract clauses which require notice of the discrepancy within a certain time limit. A charge for extra work may be waived if the contractor doesn't give notice as required by the contract. For example:

> *Contractor shall not be liable for discrepancies between representations or requirements in the plans or specifications and conditions at the jobsite unless contractor knowingly fails to report a discrepancy, in which case contractor shall be liable for additional costs incurred as a result of failure to give prompt notification.*

Even when a contract doesn't include a clause requiring timely notice, many courts imply a duty to give notice when something unexpected is discovered on the site.

Assuming timely notice, an owner will usually be liable for extra work when the contractor discovers a concealed structure, water, power, waste, drain or gas line that is inconsistent with what the plans show. Except in the case of an emergency, the contractor has an obligation to give notice before the structure or utility line is disturbed or damaged.

Differing Site Conditions

Many surprises result from differing site conditions — something on the site is different from what was assumed. Examples include unsuitable soil, debris buried below the surface, utility lines not where expected, electric or plumbing lines discovered in a wall cavity, etc. The list of possible surprises is infinite. Every CM contractor has an obligation to both reduce the list of surprises and mitigate the loss when there's an expensive surprise.

A clause on differing site conditions, such as F.A.R. § 52.236-2, is considered a benefit to both the contractor and the owner. A contractor can bid the job based on what is known and expected, not on the worst possible contingency. If site conditions differ from what was expected, the contractor will get paid for work actually done. Owners get more competitive bids with smaller contingency allowances.

> **Differing Site Conditions Defined**
>
> Federal Acquisition Regulations § 52.236-2 requires this clause in Federal construction contracts.
>
> *(a) The Contractor shall promptly, and before the conditions are disturbed, give a written notice to the Contracting Officer of —*
> *(1) Subsurface or latent physical conditions at the site which differ materially from those indicated in this contract; or*
> *(2) Unknown physical conditions at the site, of an unusual nature, which differ materially from those ordinarily encountered and generally recognized as inherent in work of the character provided for in the contract.*
> *(b) The Contracting Officer shall investigate the site conditions promptly after receiving the notice. If the conditions do materially so differ and cause an increase or decrease in the Contractor's cost of, or the time required for, performing any part of the work under this contract, whether or not changed as a result of the conditions, an equitable adjustment shall be made under this clause and the contract modified in writing accordingly.*
> *(c) No request by the Contractor for an equitable adjustment to the contract under this clause shall be allowed, unless the Contractor has given the written notice required; provided, that the time prescribed in paragraph (a) of this clause for giving written notice may be extended by the Contracting Officer.*
> *(d) No request by the Contractor for an equitable adjustment to the contract for differing site conditions shall be allowed if made after final payment under this contract.*

Courts in some states recognize two types of differing site conditions:

1. Type I is any hidden condition materially different from what a contractor is entitled to expect based on a reading of the plans and specifications. For example, a septic tank discovered on-site by the contractor would be a Type I differing site condition.

2. Type II is a hidden physical condition consistent with the plans and specifications, but very different from anything normally encountered. For example, suppose a buried septic tank appears on the site plan. But when excavation begins, the contractor discovers that the tank is filled with silt. That would be a Type II differing site condition.

A claim for extra work for both Types I and II conditions will be harder to prove if:

➢ the owner has disclosed everything known about the condition,
➢ the contractor hasn't visited the site or didn't investigate all information available,
➢ a reasonably prudent contractor would have anticipated the conditions actually found.

Federal Acquisition Regulations § 52.236-2 provides extra payments for both Type I and Type II differing site conditions. But you could limit extra payments to either Type I or Type II differing site conditions. For example, the contract could limit extra pay to Type I conditions: anything not shown on the plans. That's easy to document. Type II conditions are more difficult to prove. An expert will be required to describe what a contractor could reasonably expect. It's also legitimate to cap extra payments for differing site conditions to a specific amount. The contractor would be required to cover any expense over that amount.

The courts in many U.S. states and courts in Europe don't distinguish between Type I and Type II differing site conditions. They follow a simpler standard. The contractor gets paid for *unforeseeable* conditions. Anything that wouldn't be reasonably anticipated by an experienced contractor is considered extra work.

"Differing sight conditions? Not me! My vision is 20/20."

To reduce the opportunity for surprise, consider destructive testing. For example, cut a hole in the wall, roof or floor. The cost of repairing the damage may be less than the potential for construction claims. Otherwise, it's pot luck as to what's behind all those walls and roofs. There will be surprises. If the owner won't allow destructive testing, consider setting up a contingency fund against which both the owner and contractor can draw. When the job is done, anything remaining in the fund can be split between the owner and the contractor.

Other Differing Site Condition Issues

Abandoned work. When differing site conditions are discovered, the owner may decide on a change in the work. If the change requires that some completed work be abandoned, the owner gets no benefit from that work and would prefer not to pay for it. The contract can require that the contractor carry the risk of abandoned work.

Man-made conditions. Most differing site conditions will be physical conditions, such as concealed rock or unstable ground. But man-made conditions, such as a buried septic tank or an old concrete foundation or roadway, may also be considered a differing site condition. The contract could limit extra payments for man-made differing site conditions.

Forces of nature. Conditions such as unusually severe weather, rising water or seismic activity aren't usually considered differing site conditions. Unless the contract provides otherwise, these conditions won't be the basis of a claim for extra work.

Changes in the law. Any change in plans or specifications necessary to conform to law, code, ordinance or regulation isn't a differing site condition. But the contractor will have a claim for extra work if the plans or specs require a change. For example, once excavation begins, the building department might require the engineer to plan foundations using lower values for soil load-bearing capacity. I've even seen a building department decide to enforce the following year's code even though the county board of supervisors hadn't adopted that code yet.

Other Sources of Claims

Occasionally, owners have claims against a contractor. Most of these will involve defective work or callbacks. We'll cover warranty issues in the next chapter. Claims can also take the form of:

> chargebacks, such as for failure to clean the site, or for unfinished or unacceptable work

> damage done to existing site improvements. Examples: ripped computer cabling or food coolers shut down

> minor variation from the plans that can be accepted by the owner

> liquidated damages for late completion

> contract charges, such as for excessive requests or review of submittals

> failure of the contractor to diligently progress with the work

Sometimes a third party, such as an adjacent property owner, will have a complaint or threaten to file a claim. Regardless of the source, CM contractors will have a role in resolving the claim: investigate, notify, assess, mitigate, report and settle. To settle claims, I've had to resurface a neighbor's driveway, paint home exteriors, and clean dust out of air conditioning systems.

Resolving Contractor Claims

Everyone who makes a career of construction work will eventually get into a dispute that can't be resolved by compromise and negotiation. Personally, I feel a sense of failure when I have to get help from others (such as an arbitrator or a court) to resolve a dispute. That's especially true when I'm working as a CM contractor. Every CM contractor's responsibility is to resolve disputes without the intervention of others whenever possible.

I explained my philosophy on dispute settlement in Chapter 7. I won't repeat here what was said there. It's enough to observe that the best way to settle disputes is at once and on the spot. Two experienced construction pros eager to get on with the work understand best how to cut a deal — certainly better than two highly-compensated law firms circling each other over the spoils.

But eventually you're going to have a dispute that can't be resolved by negotiation. When that happens, some steps should be routine:

The contractor should continue working. Unresolved claims or disputes aren't a reason to suspend the work and aren't a reason to stop making payments. Continuing the work doesn't usually waive anyone's rights. Work can proceed while resolving disputes about extra pay, extensions of the time or construction defects. Find a way to keep the job moving. Don't let the construction site turn into a war zone.

Claims shouldn't be a surprise. The owner deserves fair warning that a claim is pending. The contractor's notice should include the date the right to make a claim was discovered, and circumstances that support the claim. Appeal to a mediator or arbitrator or the filing of suit doesn't relieve a contractor of the obligation to provide facts about the claim.

The contractor has to prove the claim. The contractor must have records that document both the source of the dispute and the amount claimed. If the claim involves extra work, the contractor has to provide detailed records which show each expense, including payroll records and receipts for subcontracted work, materials and equipment. It's best if these records are made available to the owner as work is being performed. If the claim is for additional time, timesheets and revised schedules will be required. Normally, no extension of time is warranted for extra work that doesn't affect the critical path. Likewise, no extension of time is required for concurrent delays — one delay occurring within the same time frame as another delay.

There has to be a cut-off point. There must be a set time beyond which claims won't be considered. That cut-off point could be final completion. But it's better if the cut-off date falls a week or two after work on the alleged change is completed.

Oral consent isn't enough. Avoid disputes that begin "I said…" or "You said…" We all have a very convenient memory. It's easy to forget what we don't want to remember. Get it in writing. No claim is settled until a written change order is approved and issued. You could also require that a certificate of accuracy accompany the final statement of claim:

> *I certify that this claim is made in good faith; that supporting information is accurate and complete to the best of my knowledge and belief; that the amount requested accurately reflects the contract adjustment for which the owner is liable; and that I am duly authorized to certify a claim on behalf of the contractor."*

The value of claims. If the contract identifies how charges for extra work will be calculated, such as a unit pricing schedule, any claim by a contractor should follow that formula. Unit pricing schedules are common when exact quantities are unknown when the contract is let out to bid. For example, there may be an agreed cost for concrete per cubic yard, including both labor and material. If the contract doesn't mention the subject, the cost of extra work has to be negotiated, usually based on the cost of labor and materials. Generally, the cost of preparing a claim, both labor cost and the cost of expert services, has to be absorbed by the party making the claim.

Mediation, Arbitration and Litigation

If you can't resolve a dispute, litigation in either state or federal court may be necessary. But there's seldom a clear winner when contract disputes have to be resolved in a court of law. If the contract requires either mediation or arbitration (sometimes called *alternative dispute resolution*, or *ADR*), courts won't allow either party to sue until mediation or arbitration is complete. Arbitration may be reviewed by a court, but no court proceeding is allowed until arbitration is complete.

We'll consider each of these three dispute resolution procedures in order of cost, from least expensive (and quickest) to most expensive (and slowest).

Mediation

Before considering either arbitration or litigation, consider mediation, either informal (with the help of a friend or associate) or formal, by a paid and experienced construction mediator. In Chapter 7, I suggested using informal mediation to settle disputes. For example, the contract could provide that disputes be referred to a contract claims committee before any further action is allowed. Informal mediation is the obvious first choice, assuming you have the consent and cooperation of the contractor making the claim. If informal mediation isn't possible or successful, go back to the contract. If the contract requires mediation by a specific organization, such as the American Arbitration Association (AAA), you'll have to follow the terms in the contract.

Mediation isn't binding on anyone. It's a form of negotiation. What you get is an informed opinion by an independent expert on how the dispute could be resolved. Even if the mediator's recommendation isn't adopted in full, mediation can serve a useful purpose:

> ➢ It's less adversarial than either arbitration or litigation.
> ➢ It forces both parties to organize their facts and prepare their best case.
> ➢ It requires both parties to listen carefully to the other side.
> ➢ It should help narrow the issues to what's truly relevant.
> ➢ It may settle parts of the dispute, even if it isn't totally successful.

Mediation, as practiced by most professional mediators, is very different from either arbitration or litigation. Mediators actively promote resolution. Most mediators will ask for authority to research the facts, interview witnesses and research the law. Arbitrators and judges don't do that. They decide the case based on the facts presented. Negotiation is left to the parties and their representatives (usually attorneys). Because mediation is so different from what you might expect, and because most contractors have very little experience with mediation, we'll offer more detail on this subject.

"You're going to settle this dispute faster than a speeding what?"

Starting Mediation

A web search will turn up at least several professional mediators in your community. Some are mobile, and will come to your construction site at a time convenient for the parties to the claim. Some specialize in settling construction claims. Try to select a mediator with unquestioned credentials and objectivity. Mediation isn't going to work unless both parties have confidence in the mediator. Any mediator you interview should be able to supply a resume. Disqualify any mediator who has a previous connection with anyone involved in the project. Before committing to mediation, get a written agreement on the fee and how that fee will be paid.

Mediators usually begin with a full explanation of their rules. Most mediators aren't lawyers and don't follow the rules of evidence. Statements and documents that couldn't be introduced in evidence in a court of law will be welcome in mediation.

The essence of mediation is compromise. The mediator will usually sit at the head of the table and ask each party to present their side of the facts. The mediator will probably insist that the opposing party sit in silence when the other side is speaking. After these presentations, the mediator will usually ask for a conference in private with each party. This gives each side a chance to refute arguments made by the other side. It's also an opportunity for each side to confide in the mediator, such as by suggesting options they're not eager to reveal to the other side. These conferences also allow the mediator to:

> ➤ point out weaknesses or flaws in their case,
> ➤ emphasize strengths in the opponent's case,
> ➤ explore where compromise is possible,
> ➤ identify key issues that require investigation,
> ➤ describe the consequences likely if there's no compromise.

The mediator will try to:

> diffuse animosity and hostility,
>
> engender a spirit of cooperation,
>
> develop a sense that progress is possible,
>
> set a timetable for resolving the dispute.

Mediators commonly ask for authorization to do whatever additional investigation is warranted. The mediator will probably offer to submit a proposed settlement when mediation is complete. If both parties agree to that offer, the mediator will have several days (or weeks) to investigate the facts and prepare an estimate of the amount likely to be awarded if the claim were to go into arbitration or litigation. This recommendation is only opinion. It shouldn't be used as a basis for further negotiation. The mediator will usually ask for access to all relevant documents and for authority to speak with everyone who has information about the claim.

Finding Common Ground

Between the first and second meeting, most mediators will consult separately with both sides, interview potential witnesses and, if necessary, read court cases that could serve as precedent in resolving the present dispute. At the end of the investigation period, the mediator will ask once again to meet with each side separately. During each of these conferences, the mediator will usually play the devil's advocate, pointing out strengths in the opponent's case and weaknesses in the party's case. If the mediator is successful, each side will recognize the advantage of compromise and settlement.

Also as part of these conferences, the mediator will usually ask for a range of values within which each party would agree to settle. The mediator may or may not reveal that range of values to the opposing party. More likely, the mediator will ask each party, "What would you say if the other side offered to settle for $X?" If the settlement numbers aren't close, no settlement may be possible. If the settlement numbers offered by each of the two sides overlap or coincide with the recommendation of the mediator, settlement may be easy.

A major mediation case could take a month or more. Most will be complete in a week or two. What the parties do with the mediator's recommendation is their own business. At the very least, both sides will better understand the strengths and weaknesses of their case and will be better prepared for either arbitration or litigation.

Arbitration

The two major arbitration organizations are the AAA (American Arbitration Association) and CDRS (Construction Dispute Resolution Services). Both offer expedited arbitration for construction disputes, including arbitration for cases under $10,000 by submission of documents. Both AAA and CDRS are easy to find on the Web.

The contract may require use of arbitration (not the court system) to resolve disputes. Generally, courts won't allow suit on a contract that requires arbitration. In the absence of fraud, a decision in arbitration will be considered final and won't be subject to court review. If the contract requires arbitration, the contract probably requires arbitration by either the AAA or CDRS. Generally, parties to the arbitration pay their own expenses. But arbitrators may have the discretion to award reimbursement of their fees.

Some states require arbitration of certain types of contract disputes. For example, the New York Construction Contracts Act requires expedited binding arbitration by the AAA of payment disputes on larger private jobs. Massachusetts allows homeowners to insist on arbitration approved by the state to settle disputes about home improvement work. Some states require a notice in the contract if disputes have to be settled by arbitration rather than in a court of law. For example, Nebraska requires a notice in capitalized and underlined type adjacent to the signature block when arbitration is the sole remedy for dispute resolution.

The advantages of arbitration include both lower cost and a quicker decision than in a court of law. Parties to the dispute select the arbitrator or arbitrators. The decision of the arbitrator has the same effect as a court decision. But there are some disadvantages to alternative dispute resolution:

- In some states, arbitrators can't require witnesses to testify or produce documents.
- There's no right to a trial by jury.
- Laws in some states and federal statutes place limits on ADR.
- Arbitrators have only limited authority to consolidate separate cases and join parties who aren't required by contract to arbitrate.
- ADR can't be used to enforce mechanic's liens.
- There's no effective way to appeal an arbitrator's decision, even if it's completely wrong.

Other points to consider if your contract requires arbitration:

- *Where should arbitration be conducted?* You probably want the arbitration hearings held in the county where the job is located.
- *What rules should the arbitrator follow?* Arbitration associations have rules which apply to arbitrations conducted by their members unless the parties select other rules.
- *Should arbitration be required of everyone on the job?* ADR may be more effective if all architects, engineers, contractors and subcontractors on the job have signed similar arbitration agreements. That may allow the arbitrator to consolidate all claims into a single hearing.

> *Can the arbitrator award costs and attorney fees?* Arbitration rules usually provide that parties pay their own expenses in arbitration and pay only their share of the arbitrator's fees. But you may want to give the arbitrator authority to award reimbursement of fees to the prevailing party.

> *Is small claims court a better option?* Most states allow use of small claims court if the amount in dispute is under some set amount, usually between $2,000 and $10,000. Small claims court will nearly always cost less than arbitration.

If arbitration or a legal action is required to resolve a construction dispute, a CM contractor will have an obligation to prepare the documentation, attend conferences, interview witnesses, offer testimony and present the owner's case. Of course, the role changes if the CM contractor is the party making the claim.

Litigation

Without an agreement to arbitrate, disputes have to be settled in court. Courts in each state have their own rules and procedures. But you can still make some decisions for the court:

> *Where will disputes be resolved?* Nearly all states require that disputes about construction in the state be resolved in the state. But you may be able to decide on the court within that state where the matter will be heard.

> *Will the court have authority to award reimbursement of fees and costs?* The threat of an award of legal fees tends to discourage litigation of weak cases. Most states won't award reimbursement of legal fees and court costs unless required by contract.

Be aware that courts are in the deciding business, *not* the collections business. A court judgment is like a hunting license. You still have to find assets and execute against those assets. Getting a judgment doesn't mean you're going to collect anything. Before you file suit, consider doing a search for assets to make sure the defendant can pay. The losing party may not *have* any assets, so even if you win, your winnings may be zero — you'll be out both the amount awarded by the court *and* your legal fees.

Nearly Done

We're getting near the end of the project. Work is winding up. The owner's anxious to begin using the building. Only a few (important) points remain: Project closeout, working off the punch list and warranty. Those are topics for Chapter 13.

Chapter 13

Directing Project Closeout

IN CHAPTER 10, WE covered suspension and termination of the job before completion. Assuming the job isn't suspended or terminated, a CM contractor's last major task will be project closeout: the owner's walk-through, acknowledgment of substantial completion, the punch list, final completion, callbacks and warranty. That's the agenda for this chapter. Final payment, release of retainage, and lien waivers are logically part of project closeout. But we covered those issues in Chapter 9, *Evaluating Payment Requests*.

Before getting into completion, we should touch on another possibility: early occupancy.

Beneficial Occupancy

Early occupancy (sometimes called *beneficial occupancy*) can be both a major advantage to the owner and a significant distraction to the contractor. A contractor who allows beneficial occupancy is giving the owner permission to begin using all or part of a building before acknowledging substantial completion. Early occupancy always raises insurance, liability and operating expense issues. If these issues weren't covered and resolved in the contract, they should be settled before reaching any agreement on early occupancy.

On most residential and commercial jobs, occupancy will require final inspection and a certificate of occupancy from the local building department. Without that certificate of occupancy, utility companies usually won't start service. That keeps most unfinished residential buildings uninhabitable for normal purposes. But "occupancy" doesn't necessarily mean that people are living in the building and using utility services. Occupancy by the owner could include storage of vehicles or personal possessions in or near a building still under construction.

Early occupancy can be a contentious issue if the owner wants to begin using part of the building but doesn't want to acknowledge substantial completion of the portion occupied. There are at least four ways to handle beneficial occupancy in a construction contract:

Case 1. The owner can take up occupancy of part of the work at any time and isn't required to acknowledge partial substantial completion.

Case 2. The owner can occupy part of the work when 90 percent of the contract price has been paid. No acknowledgment of partial substantial completion is required.

Case 3. The owner can occupy substantially complete parts of the work. Acknowledging substantial completion requires payment for everything in that part of the work except what's on the punch list.

Case 4. Early occupancy constitutes acceptance. If the owner occupies or uses any part of the work before substantial completion, the portion occupied is considered finished, accepted and the responsibility of the owner.

Early occupancy can also raise insurance, liability and operating expenses. The contractor may ask for compensation for any delay or extra expense due to early occupancy. Even if the owner is only storing furnishings or vehicles on-site, it's still reasonable to:

➢ get prior consent,
➢ define the area to be occupied,
➢ identify paths of travel to and from the occupied premises,
➢ provide proof of general liability and fire insurance coverage,
➢ ask the owner to waive any claim for loss or damage,
➢ inspect the area and create a local punch list before the owner goes into possession.

An owner eager for early occupancy may be willing to negotiate terms favorable to the contractor, such as either waiving liability for late completion or gaining a bonus for early completion. But early occupancy shouldn't relieve a contractor of the obligation to complete the work and shouldn't require compromise of any claim an owner has against the contractor.

On large projects, early occupancy usually happens on a room-by-room basis. On a school job, for example, the contractor might turn over classrooms for occupancy, but not the shop spaces.

Points to consider before granting early occupancy:

"It's all ready to move into, except for the doors and windows. They should be delivered in 6 to 8 weeks."

➢ Nothing is ready for beneficial occupancy until fire walls are complete, fire alarms are installed, tested, inspected, active and under permit.

➢ A punch list is necessary for each portion of the project before early occupancy begins. Completing punch list items is always more difficult with early occupancy. On a commercial building, the punch list may have to be completed after normal working hours. That can require overtime pay at higher rates.

➢ Getting people in and out of occupied portions of the project can be a problem — driving the insurance guys a little crazy. OSHA isn't too crazy about the idea of early occupancy either. Make sure all walkways are direct, obvious and unobstructed. Post signs and fence off the rest of the construction site.

➢ If the warranty period starts at occupancy, partial occupancy will result in multiple warranty periods, all running concurrently. Send a memo to the owner noting the date when each warranty period begins.

Inspection for Substantial Completion

A project is a candidate for substantial completion when the building can be used for the intended purpose. The contractor should notify you when it's time for the walk-through inspection. You set the time and date for the inspection. The contractor, owner and the designer should join you for the walk-through. Everyone will need a pencil and a tablet to make notes. You'll consolidate these separate lists into a single punch list of items to be completed after occupancy. But review the owner's notes. The owner's punch list shouldn't turn into the owner's wish list. The contractor is obligated to complete only what's in the contract.

This inspection will be easier if the contractor prepares a preliminary punch list of known defects and submits that list with the request for inspection. If you agree that items on the preliminary punch list are consistent with substantial completion, schedule the walk-through at a time convenient for everyone concerned.

I wouldn't consider doing the walk-through until:

- installed equipment has been tested and found to be in working condition,
- the contractor has completed performance tests required by the specs,
- reports, maintenance manuals, operation instructions, warranties, keys and control devices have been delivered, reviewed and approved,
- the owner or owner's representative has been briefed on equipment operation,
- record documents have been delivered to the owner, including changes to the contract, as-built drawings, submittals, shop drawings, test results, inspection certificates, surveys and logs,
- debris, waste, and excess materials have been removed from the site,
- signage and safety devices (i.e. fire extinguishers) comply with code requirements,
- the job has passed final inspection,
- occupancy has been approved by the public authority,
- all utilities and services are connected and operating.

Plan to terminate and reschedule the walk-through if you discover:

- anything that would limit the intended use, such as failure to connect to the sewer main,
- anything which would be difficult to remedy while the building is occupied, such as flooring that doesn't comply with the specs,
- too many incomplete items, regardless of the type. Five incomplete minor items would be acceptable; 35 would not,
- anything significant that will require several days or weeks to complete.

> **Not a Time to Hurry**
>
> As the CM contractor, you're probably as eager as the prime contractor to finish the present job and move on to the next. Closing out the job promptly will be in your best financial interest. But granting beneficial occupancy or acknowledging substantial completion too soon can delay final completion. Assume that after occupancy, completing the punch list work will cost about 50 percent more and take about 50 percent longer than doing the same work before occupancy. Don't be in too big of a hurry to close out the job.

Acknowledging Substantial Completion

The owner has to make a decision when the walk-through is complete. Will the owner acknowledge substantial completion? Or should another walk-through be scheduled later? Acknowledging substantial completion means, of course, that most of the remaining contract price is due.

Acknowledging or rejecting substantial completion is a decision for the owner. If it's a close question, your recommendation may tip the balance. When helping the owner decide, cite considerations such as:

> *The nature and extent of work yet to complete.* You have a good perspective on how much work remains. Is the owner's occupancy of the site consistent with continuing work by construction trades?

> *Has the contractor been responsive to reasonable requests during construction?* A contractor who hasn't been cooperative when there was a heavy financial incentive to cooperate may be very difficult to deal with when that financial incentive is gone.

> *Has the contractor's A-team gone on to work elsewhere?* When only punch list work remains, the only crew left on-site may be a third-string handyman. More oversight may be required to get first-class performance. Whining on the phone about faulty work or slow progress to the contractor may actually delay work. A first-string plumber or carpenter may be available. But that plumber or carpenter is probably working fulltime on another site across town — and may only be available after 3:00 p.m. or on weekends.

> *Will acknowledging substantial completion delay final completion?* Work on-site is nearly always more difficult with an owner in possession of the premises.

> *How eager is the owner to go into occupancy?* An owner can be like a kid with a new toy — eager to try it out. But making the owner happy at this point may not be the best choice for project completion. On commercial jobs, the opening date may be cast in granite. For example, on a casino project, the blackjack dealers, bartenders, restaurant help, maids and management staff have probably been given a day to report to work. Not opening on time packs an enormous financial punch. Few contractors can afford to delay a project like that. On a Neiman-Marcus store in San Francisco a few years ago where I was the CM, opening night was a charity event with 5,000 VIPs invited. There was no leeway — the construction team either finished on time or virtually left in handcuffs for Alcatraz.

An owner with deadlines to meet may have no alternative but to acknowledge substantial completion, declare a victory, and move in. That makes the decision easy. Under those conditions, your only obligation is to explain that occupancy too soon has consequences that may not be easy to reverse.

You'll usually have plenty of advance notice if the owner is itching to get the reins and run off with the project, done or not. Finance-driven projects are notorious for time pressure. You'll hear about it from the owner at every meeting. "Why can't we move in next weekend?"

Defects, Trivial and Serious

The most difficult issue during the final walk-through will be discovery of a defect that wasn't corrected earlier, especially a defect that doesn't comply with the building code. There's no easy way around a problem like that.

"Look at the advantages: When the water gets high enough the termites all drown."

If the issues are minor, go with the flow. At the end of a job, everyone wants to finish and get on with something else. That includes the owner. The contractor has probably pulled off the site. When the contractor is there, he's conducting business over the hood of his truck in the parking lot. The owner's distracted with other issues and probably has no interest in the last 16 items on the punch list. Essentially, the owner has arrived where the contractor has been for a month. Both owner and contractor want to be done with it. As a CM contractor, I'm willing to let the owner and contractor close out work even if it's technically not done yet. To do otherwise is to risk assault from both sides.

Assuming there isn't a code issue, I abruptly end my participation. I let the owner take over the remaining punch list items. That work may never get done, but no one seems to care. By backing away, I don't become a victim of the end game. I may recommend that the owner take a credit for unfinished work. Then I go on to something else, just like the owner and the contractor have done already. If everyone's reasonably happy with the outcome, that's good enough.

Serious Defects

If the issue discovered at final walk-through is a serious code violation, there's no choice. No one is done until it's fixed.

The contractor will claim it's too late to cry foul. The work has already been approved! That may or may not be true. If the work complies with the plans and specs and if a submittal or sample was approved, I agree with the contractor. The correction will be done at extra charge. But progress payments alone aren't approval of anything. Both the owner and the building inspector have the right to reject work right up to completion of the punch list. But that's no excuse to wait until the last possible minute to point out an obvious mistake. Here's an example:

> On a $9,000,000 school shop facility, I noticed that the contractor wasn't using fire-rated caulking in penetrations through one-hour fire walls. He simply covered gaps with siding and trim. I saw this mistake and sent the contractor a memo. The state inspector got a copy of my memo and noted the problem in his report. Still, it didn't register with the contractor.
>
> The owner was eager to take possession of the facility. He had classes scheduled for the next week. On the final walk through, I made a point of asking about the fire caulking. There wasn't any. What should I do? I explained that I couldn't sign off on the job until fire caulking was installed and approved by the state inspector.
>
> The owner was grossly unhappy with my decision. The contractor was irate. He had to tear off all the interior trim work around the wall joints and install the required fire caulking — at his own expense.
>
> The work got done — but it was two weeks late. It wasn't an easy choice. But I stood my ground. My conscience was clear and no students were ever at risk. That's the higher calling — beyond doing exactly what the owner would like me to do in the heat of the moment. Safety is always job one; everything else comes in second.

If you have a state license, you're licensed for a reason. You represent the governing authority, and have an obligation to comply with the code. So, what should you do when you see a code violation at final walk-through? If it's a so-so issue, like the sidewalk slope that's 1 percent too steep, I'd email the owner and the contractor just to create an *"I told you"* record. If it's a life safety issue, like an improperly-built firewall or a missing balcony railing on the 11th floor, I'd email the owner, contractor and the inspector, advising not to open the building until the defect is corrected. If that doesn't work, I'd refuse to sign off on the punch list. And if that still doesn't get the desired result, I'd notify the building department. My license is at stake. I may only do one building with that owner, but I plan to work with the building department for decades. Anything less than following through on a safety issue would be unprofessional and foolish. I hope you agree.

Partial Completion

Acknowledging substantial completion doesn't have to be an all-or-nothing affair. If the job includes distinct parts, such as a residential duplex, you could recommend acknowledging substantial completion for one unit and not for the other. In that case, your acknowledgment should note what's accepted and what's rejected. Another choice would be to exclude an entire portion of the job (such as all electrical work) from your acknowledgment of substantial completion.

Partial completion presents a problem when equipment serves several portions of the project — some of which is complete and some incomplete. HVAC and electrical systems are obvious examples. By convention, consider as complete any equipment or system that serves a portion of the project that's acknowledged as substantially complete, even if that equipment or system also serves incomplete portions of the project. That means the owner has to pay a little more a little sooner. But there's little or no extra risk to the owner.

Here's another alternative if the owner is eager to take up occupancy. Instead of adding a defect to the punch list, have repairs done under warranty. That's usually the best choice when defective equipment is covered by a manufacturer's warranty.

Assuming the job fails the first walk-through, which most do, you'll have to repeat the process at a later date. It's OK to limit the next inspection to just the items found to be incomplete in the first inspection. Any time a job fails inspection for substantial completion, send the contractor a written list of the discrepancies found. Describe each defect, whether incomplete, defective or not in compliance with the plans. If the first walk-through wasn't completed, indicate which parts of the job weren't checked. You can threaten to deduct the cost of second and subsequent inspections from money due the contractor, but I don't like to do that. The result will be more hostility than cooperation.

After substantial completion, roles change. The owner can enter and use the site at any time, while the contractor loses the right to enter the property at will.

The Substantial Completion Punch List

Acknowledging substantial completion is an agreement by the owner that only items on the punch list remain to be completed or corrected. The punch list delivered to the contractor should show:

> A detailed description of each defect. To make corrections easy to find, walk the project with a roll of blue painters tape. Stick a short piece of tape to each defect. Write a pencil number on the piece of tape that corresponds to the defect number on the punch list.

> The work required to complete or correct each defect.

> The subcontractor or trade responsible for each defective item.

> Dates when work will begin and be finished for each item.

> The estimated cost to complete or correct each defect. You'll need this estimate to figure how much of the final payment to withhold. Generally, the owner withholds 125 percent of the value of punch list items.

The description of each defect is a product of the walk-through. The remaining items on the list (work, trade, dates and cost) should be completed by the contractor. Plan to get punch list items worked off in a week. If work drags on for more than a month, a backcharge may be required to motivate completion, especially if repeated attempts to contact the contractor have failed. Failure to complete work 30 days after the forecast completion date should give the owner the right to make the correction and backcharge the contractor for the cost, including any cost of architectural services.

Anything omitted from the punch list is usually considered accepted. Of course, the contractor may still have an obligation for callbacks and warranty claims if a defect is discovered later. But acknowledging substantial completion doesn't necessarily waive the right to have latent (hidden or unrecognized) defects corrected. Nearly all states recognize an implied warranty of workmanlike construction. Even if the contract doesn't include a written warranty, the state's implied warranty will probably require the contractor to make warranty repairs for some reasonable period. We'll discuss warranties and callbacks later in this chapter.

After Substantial Completion

The owner and the contractor should sign a memorandum acknowledging substantial completion when the project or a specific portion of the project is ready for occupancy. The acknowledgment should show the date of acceptance and the work accepted if less than the entire project. The contractor will probably want language in the memorandum acknowledging that the owner takes full responsibility for maintenance, safety, utilities, controlling access, and insurance on the site. The warranty and callback period begins running from the date on this memorandum. If your project is completed in pieces or phases, you'll need multiple acknowledgment memos.

As mentioned, roles reverse after substantial completion. But the contractor still has obligations:

> The contractor remains responsible for the crews while they're working off the punch list. For example, the safety of the contractor's crews remains the responsibility of the contractor. Any damage done by the contractor's crews should be repaired by the contractor.

> The contractor has an obligation to remove from the jobsite anything that belongs to the contractor. That includes barricades, construction tools, equipment, supplies, and all temporary structures used during construction. Be sure erosion and contamination control measures are removed from streets and drainage locations.

- The contractor should notify subcontractors, insurance carriers and sureties that construction is complete. If required in your state, file a notice of completion with the county recorder. Notice of completion starts the clock running on the lien period for all trades.

- The contractor can enter the jobsite only for the purpose of completing or correcting items on the punch list or when doing callback or warranty work requested by the owner.

- Except as provided by contract, the obligation of the contractor to maintain public liability, property damage and builder's risk insurance on the project comes to an end.

Final Completion

On a larger job, you need to authorize the designer or the inspector of record to sign off on all punch-list items. On a smaller job, the owner will be on-site and can sign off on items as the punch list is worked off. There shouldn't be any need for another walk-through when the contractor has completed all punch-list items. But the owner needs to be informed when the contractor claims final completion. If the owner agrees that punch-list items have been corrected, the CM contractor should advise the contractor to prepare and submit an invoice for final payment.

Final payment is due when:

- the punch list is complete,
- lien releases have been received from everyone with lien rights,
- both you and the owner agree that work is complete,
- contingent claims (such as for change orders) have been settled or withdrawn,
- the surety (bonding company) has given consent to final payment,
- the lending company, if there is one, has signed off on the project if required.

When these conditions have been met, recommend that the owner make final payment, including any retainage. That recommendation should be accompanied by your final status report on the project.

The project is complete when the contractor accepts final payment as resolution of all claims, except as required to meet callback or warranty obligations. The date final payment is accepted is usually taken as the date of final completion.

The Final Project Report

Back in Chapter 8, we emphasized the importance of keeping the owner informed with regular status reports. The owner is due a final status report when you present a final invoice for your fee as CM contractor. That final report should include a narrative history of the project, summarizing:

> ➢ dates of significant project milestones
>
> ➢ costs and payments
>
> ➢ start of the warranty period
>
> ➢ any issues that remain to be resolved

Include with the final project report a summary of warranty information, including both manufacturers' warranties and warranties by the contractor.

> ➢ identify the materials and equipment covered by the warranty
>
> ➢ describe the types of defects that are covered or excluded
>
> ➢ list contact numbers to call if there's a warranty claim
>
> ➢ offer to assist with processing warranty claims, usually at an hourly rate
>
> ➢ list any spare parts, matching paint or service items that are being delivered to the owner

I usually put all of this information into a three-ring binder and give the owner at least three copies: one for maintenance, one for his secretary and one for him to misplace somewhere.

Callbacks

Most contractors acknowledge responsibility to correct defects in material or workmanship discovered after work is completed. Laws in many states and many model construction contracts impose the same obligation, usually referred to as a *callback* period by professionals in the construction industry. Warranties are different. We'll cover warranties in the next section.

Most contractors distinguish between their responsibility for callbacks, and warranty claims. Callbacks require a contractor to return to the jobsite, inspect the defect and make any repairs necessary. The essence of a *warranty claim* is the right to collect money damages. Prudent owners usually prefer both warranty and callback protection.

The essence of a callback is that something wasn't installed or isn't working the way it was supposed to. Considered in that light, a callback is a repair made as an extension of the construction process. Typical callback repairs include floor squeaks, adjustments to HVAC or electric controls, doors or windows that don't work quite right, roofing or rain gear that fails during the first rain, or piping that leaks for no apparent reason. Many callbacks are items that could have been discovered with a more-thorough walk-through inspection. Others result from premature failure due to faulty materials or workmanship.

Some construction contracts provide a callback period that runs for a specific period, from as little as 30 days to as much as a year. Other contracts offer callback protection only for the period between substantial completion (when the punch list is developed) and final completion (when the punch list has been worked off). On some jobs, substantial completion and final completion may happen at the same time, resulting in no callback period. After the callback period expires, claims can be handled under the warranty.

If the first callback doesn't correct the problem, a second will be required. A new callback period begins running again after the first callback, but there has to be some reasonable limit to the callback repair period.

Here are other generally-accepted rules on callbacks:

> Failure of the owner to give notice of a defect within the callback period is a waiver of the right to repair or replacement.

> If the contractor fails to respond to a callback notice within a reasonable time, an owner has the right to make repairs at the expense of the contractor.

> A contractor has the right to test and inspect any claimed defect during the callback period. If required, the contractor can get the opinion of an independent expert before beginning repairs.

> The callback period starts early for portions of work that the owner occupies early.

Express and Implied Warranties

Warranties against construction defects can be express (written in the contract) or implied, either by court decisions or by statute (state law). All states enforce the terms of an express warranty in the contract. Most states also require that residential construction be done in a workmanlike manner. That's an implied warranty. It's not part of the contract and will be entirely separate from any written express warranty.

Many states restrict or limit any attempt to disclaim this implied warranty of workmanlike construction. For example, in a contract for residential construction, Kansas imposes a fine of up to $10,000 for trying to disclaim the implied warranty of fitness for purpose. Other states permit a contractor to disclaim implied warranties if the disclaimer meets specific requirements — type size, wording, placement in the contract, etc.

The term and coverage of implied warranties will be whatever a court decides on a case-by-case basis, and may vary from state to state. For example:

Minnesota's statutory warranty for residential work has to be written into every contract — three paragraphs of very precise language. The warranty runs for one year on materials and workmanship, two years on plumbing, electrical and HVAC work, and 10 years on any "major construction defect."

California makes residential builders liable for a long list of construction defects. The builder has to either make repairs or compensate the owner for the loss — including relocation and storage expense. The warranty may expire in one, four or 10 years, unless another expiration period is defined.

Pennsylvania law implies a warranty of good workmanship; what's *reasonable* under the circumstances, *not* perfection. An implied warranty of habitability is breached if a defect presents a "major impediment to habitation," an issue for courts to decide. Both residential and non-residential construction are covered by the Pennsylvania warranty.

New Jersey provides an express Home Owners Warranty (HOW) to all buyers of new homes. Every new home sold in New Jersey comes with a limited warranty against construction defects for up to 10 years. Home builders in New Jersey are required to enroll in either the state warranty plan or a private warranty plan approved by the state.

Express Warranties

Written warranties in a construction contract can be:

> *On named components.* This warranty is like one on a new car. Defects covered by warranty are listed one by one in the agreement: Concrete, framing, wallboard, etc. The warranty period is defined for each type of material. Anything not specifically identified isn't covered by warranty.

> *On all materials and workmanship.* The contractor warrants that the work will be free of defects due to faulty material or workmanship for the period specified in the agreement.

> *A broad form warranty.* This is the most common warranty in public works construction and is required on nearly all projects funded by the federal government. The contractor warrants that the work will be free of defects in material, or design furnished, or workmanship performed by the contractor or any subcontractor or material supplier, for the period identified in the agreement — usually one year.

An express warranty can run for any length of time. Warranties on cosmetic defects usually run for a few months to a year. Warranties of plumbing, HVAC and electrical work usually run from two to four years. Warranties on structural portions of the building commonly run for five to 10 years. Any failure of a load-bearing portion of a building that compromises safety, habitability or sanitation would be a breach of the warranty on structural components.

Warranties usually begin running from acknowledgment of substantial completion. For anything on the punch list, the starting point will usually be the day that item was signed off on the punch list. The most favorable option for an owner would be to start the warranty running at final completion. That's usually the day the contractor accepts final payment. The contract could also provide that a new warranty period begins running when warranty repairs are completed.

I prefer to start the warranty period from the moment the punch list is complete, whether it's a room or the entire building. I never run warranty time periods on individual systems or separate pieces of equipment. That's too complex for me.

A contractor's warranty is entirely separate from manufacturers' warranties common on most electrical and HVAC equipment. For example, most water heaters come with a warranty from the manufacturer. The contractor doesn't have any responsibility for failure of the water heater during the warranty period. But the contractor does have an obligation to pass warranty information on to the owner before substantial completion. It's common practice to ask the contractor for help when needed to enforce terms of a manufacturer's warranty.

Roofing material manufacturers usually provide a warranty on the roof. An on-site inspection may be required before the warranty goes into effect. Most roof manufacturer warranties state that after 10 years, the owner recovers only half the value of the roofing material.

As with callbacks, failure to make a claim within the warranty period is a waiver of the right to repair or replacement under warranty. If a claim is made within the warranty period, the contractor has the right to test and inspect anything claimed to be defective. If necessary, both you and the contractor can get an opinion from an independent expert.

No contractor is eager to deal with warranty issues. Some contractors make it their practice to delay response to warranty claims until the warranty period has run. But you've got leverage if you care to use it. As a CM contractor, you're in a much better position than the owner to make life unpleasant for an uncooperative contractor. If you're not currently working another project with the contractor in question, you may have future projects with that contractor, so he'll want to keep you happy. Make it clear that you won't deal with a contractor who doesn't honor legitimate warranty claims. In states where contractors are licensed, consider sending the state licensing board a complaint letter or two. That should put an end to any resistance in short order.

Many states prohibit suit over residential warranty claims until the parties have followed a settlement procedure established by state law. Usually the contractor has 30 days to respond to a warranty claim. If the claim isn't settled, state law may require use of a state-sanctioned dispute resolution procedure. If that process doesn't settle the matter, the owner is then permitted to use the court system.

Warranty Exclusions

Most express warranties exclude coverage of specific types of damage and specific defects. For example, it's common to see many of the following excluded from coverage in an express written warranty. Likewise, most of the following won't be covered by an implied or statutory warranty.

- Defects in detached buildings or improvements that aren't a part of the main structure itself. For example, if the main building is a residence, a detached carport or detached garage might not be covered by warranty.
- Losses that are a consequence of a defect but aren't a physical defect. For example, a warranty probably doesn't offer protection if a building becomes uninhabitable due to the presence of toxic or hazardous materials, mold, radon gas, or other pollutants or contaminants.
- Damage to personal property or bodily injury.
- Damage resulting from circumstances beyond the contractor's control. For example, warranties offer no protection against accidents, fire, explosion, smoke, falling objects, wind-driven water or an unforeseeable change in the underground water table.
- Damage covered by insurance.
- Damage resulting from soil movement. Everything shifts and subsides. We live on drifting continental plates.
- Insect or animal damage.
- Dampness or condensation that results from improper ventilation, leaky sewer lines or pipes that burst in cold weather.
- Damage that results from use for other than the intended purposes, such as closing off exits required by the code.
- Damage caused by overloading beyond the design capacity, like a hot tub on a cantilevered deck.
- Defects that are obvious and accepted at the time of completion.
- Work completed or altered by the owner or other contractors.
- Negligence, improper operations or improper maintenance, such as closing off return air vents or failure to change air conditioning filters.

- Failure to comply with the manufacturer's warranty instructions, such as storing flammables next to a gas furnace.
- Failure to give notice of a defect within a reasonable time.
- Modifications made to building components, such as cutting a 2-inch notch in the bottom of a wood beam.
- Failure to take timely action to mitigate damages.

Warranties never cover normal wear and tear. I've had owners ask me to get the contractor back out to repaint walls that were clearly damaged by the owner's moving crew. I've also had an owner call a contractor back six months after completion when the owner's staff spilled something on the carpet. Those aren't warranty issues. They're maintenance issues. You just have to tell that to the owner.

It's common to have a construction defect cause other damage. For example, a leak in a roof caused by a construction defect can result in water damage to the building interior. In that case, is the contractor (or the roofer) obligated to replace carpet and flooring? Courts will use the law of negligence to allocate liability for casualty losses. How far does liability extend down a chain of consequences? If the damage was reasonably foreseeable, the contractor may be liable. The construction contract could settle this issue, requiring the contractor to accept liability for consequential damages caused by a breach of warranty.

"It's common to have a construction defect cause other damage."

The most common warranty claim by far is water damage. I've had a roofer fix leaks a dozen times on the same building and still have more leaking. There are literally dozens of reasons why a roof will leak, such as someone stepping on a stray nail on the roof, puncturing the membrane. If someone else picks up the nail, there's no way to know the roof was damaged until the rains arrive in February. Water may then migrate sideways through the membrane until it finds a way into the interior, and the puncture that's over the living room causes a leak in the kitchen.

To avoid callbacks and warranty claims on the roof and rain gear, order a water test.

Responsibility for Warranty Claims

During construction, communication between the contractor and the owner is through the CM contractor. That changes at project closeout. Most warranty claims are made long after construction is complete — and long after the CM contractor has gone on to other work. During the warranty period, it's better to encourage the owner to work directly with the contractors and manufacturers obligated to respond to warranty claims. Still, the CM contractor should:

> Provide a list of contacts who can respond to warranty claims.

> Offer to assist the owner in resolving warranty claims, if needed, on an hourly basis.

> Agree to monitor any required remedial work, also on an hourly basis.

What's Next?

We've described what we see as opportunities for construction managers. The next step is up to you. To begin developing a satisfying career as a CM contractor, we recommend downloading the free construction management contracts at http://PaperContracting.com.

INDEX

A

A/E team
    CM selecting 57-60
    cost issues..72
    keeping job moving 137-139
    owner selection 61
Accidents, jobsite 146
Accountant, tracking expenses........ 192
Accounting, payment process 172
ACI Manual of Practice 124
ACORD Corporation 128
Act of God delays 139-140
Addenda, Invitation to Bid 96
Agency CM contract 34
AIA model contracts 119
AISC Manual of Steel
    Construction 124
AITC Timber Construction
    Manual.. 124
American Arbitration Association
    arbitration 243
    mediation 241
American Concrete Institute (ACI)..124
American Institute of Architects
    (AIA) 119
American Institute of Steel
    Construction (AISC) 124
American Institute of Timber
    Construction (AITC) 124
American National Standards
    Institute (ANSI) 124
American Society for Testing
    and Materials (ASTM) 124
American Society of Heating,
    Refrigerating and Air Conditioning
    Engineers (ASHRAE)...................... 124
Approval items
    architect's review 162
    procedures 160
    tracking submittals 161-162
Arbitration
    advantages...................................... 244
    claims and disputes...................... 243

Architect
    errors and omissions insurance...232
    project development................. 38-39
    scope of work 62-63
    selecting A/E team 57-60
    submittal review 162
Asbestos.. 150
ASHRAE 124
ASTM 124
    master data numbers.................... 199
At risk CM contract.............................. 20
Auditor, public works payments 190
Awarding contract...................... 100-101

B

Backcharges.. 187
Beneficial occupancy 247
Bias, contract.. 116
    model contracts............................. 119
    terms that shift bias 117
Bid
    addenda 96
    comparisons.................................... 101
    errors.. 101
    invitation.. 95
    opening.................................... 100-101
    pre-bid conference 97
    regulations 95
    supplier 121-124
Bid bond 130
Bid competitions 87, 89
Bid form.................... 101-102, 107-108
Bid package, open bidding 89
Bill paying, CM contractor role......... 14
Blasting materials............................... 150
Bonds 130-131
    dispute resolution......................... 130
Bookkeeping, tracking payments ...190
Brand name specs 205
Brand names, specifying.................... 72
Broad interpretation, plans.............. 202
Budget, based on incomplete
    plans 206

Builder's risk insurance 129
Building Code Compliance for
    Contractors & Inspectors..................... 81
Building department, permit
    application.................................. 80-82
Building permit fee 81

C

Callbacks
    contractor repairs.......................... 257
    period of protection...................... 258
Cardinal change doctrine................. 219
Case study.. 23-31
Cash payments, avoid 33
Certificate of insurance 128
Certificate of occupancy................... 248
Certified payroll 190
Chain of authority.............................. 207
Change Order Agreement................ 222
Change orders
    contract amendment..................... 223
    contract details 115
    extend completion date 219-220
    force account................................. 215
    markup.. 219
    mutual agreement......................... 215
    negotiating price 216
    payment process 174-175
    processing 221, 223
    reasons for...................................... 213
    ROM estimate................................ 221
    time and material cost.......... 217-218
    value engineering 224
Channels, information.............. 157-158
Chargebacks, owner claims............. 239
Checklist, plan review 81
Church jobs 43
Claims
    against architect 232
    arbitration 244
    change order.................................. 224
    collecting information.................. 229
    construction surprises.................. 234

contractor 232-233
   delay, avoiding 143-144
   errors in plans............................235
   extra work 233-234
   liquidated damages 144
   litigation 245
   mediating 241
   mitigating loss 230-231
   owner-caused delay 140-143
   owner-generated 239
   planning settlement.....................232
   reducing risk..............................227
   release at final completion... 178-179
   resolving 239-340
   site disclosures 235
   warranty 262-263
Cleanup 149
Clients
   finding 40-41
   making an offer 47-48
   qualifying..................................41-42
   selling your services 43, 44-46
CM contractor
   agency CM contract........................34
   approving schedule................136-137
   at risk contract..............................20
   avoiding disputes 151
   bonds and insurance 127-131
   contracts 13, 112
   dealing with suppliers 25-26
   design involvement 38-39
   directing job closeout 247
   documentation for claims............245
   early project involvement..............37
   evaluating payment requests......172
   evaluating projects.........................44
   face time with owner............ 192-193
   fee.......................................14, 17-19
   final project report 257
   finding prospective clients 40-41
   hourly rate 17
   how to sell services.........................43
   identifying defective work 147
   importance of contract... 24-25, 30-31
   job conference..............................208
   job log 165-166
   jobsite cleanup............................150
   keep information flowing............154
   keep role clear 27, 29
   licensing 14-15
   limits of authority 196-197
   lump sum fee.................................18
   making an offer 47-48
   monitor safety program...............146
   negotiating change orders....216-217
   notice of claim 229
   ordering materials 121
   owner's advocate...........................39
   percent of job cost fee 18-19
   preventing confusion in role 23
   processing change orders............214
   profit potential 16-17
   project schedule 82
   recommend invoice payments....182
   record documents 168
   reducing risk of claims................227
   rejecting invoice payments...183-187
   resolving claims 239-240
   responsibilities 9-10
   retaining records 163
   role in construction.... 52-54, 125-127
   role in payment process.........14, 171
   scheduling....................... 84, 131-135
   selecting A/E team 57-60
   selling services 43, 44-46
   setting design limits 67
   successful sales approach 50-51
   tracking payments 189
   trades, scheduling..........................84
   walk-through inspection 250
   warranty law 16
   weekly/monthly charge................18
   weekly project estimates.......... 78-79
Code, intent of 204
Code compliance, check for............68
Collapse insurance.........................128
Commercial project, CM
   contractor role........................ 10-11
Communication
   contact list 167
   correspondence file.....................163
   importance of 24-25
   job conference..................... 207-208
   set policy for job............................63
   suppliers......................................195
Competitive bidding 87, 89
   regulations 95
   supplier 121
Completion date
   change order to extend 219-220
   scheduling....................................84
Comprehensive general liability
   insurance 128
Consensus DOCS model contracts ..119
Conservation, LEED 144-145
   labor and materials.....................228
Construction, CM contractor role.. 52-54
Construction consultant
   CM contractor fee 14, 17-19
   CM contractor services 10
   importance of contract 30-31
   role in construction................125-127
   working with 6
Construction Contract Writer
   CM contract options......................65
   contract software...........................56
   pre-construction contracts 64-65
Construction Dispute Resolution
   Services 243
Construction management
   contracts 112
   contracting 9
   owner's representative 8
Construction publications, project
   listing..95
Construction surprises, claims.. 233-234
Construction trades, scheduling.......84
Consultant, construction....................6
Contact list 167
Contemporaneous business
   records...164
Contingency time..........................135
Contract documents 96
Contract, awarding 100-101
Contracting
   changes over years 7-8
   paper, defined............................ 5-6
   paper, legalities 6-7
Contractor, trade, contract to owner...13
Contractors, general
   bids..87
   bonding requirements.................127
   change order claims....................224
   claims................................. 232-233
   claims for delay 140-143
   claims for extra work 237-238
   contract terms that shift bias....... 117
   compensation for early
      occupancy 248
   cost estimate 77-78
   defective work............................148
   ECI 37-38
   final completion invoice 256
   hazardous materials 149
   insurance requirements 127
   jobsite safety 146
   lien rights 187
   preliminary punch list 250
   pricing change orders.......... 215-216
   project development................ 38-39
   qualifying for bid 89-90
   reporting plan discrepancies.......236
   role in construction................125-127
   statutory employer 27
   termination for cause 211
   traditional role..........................7, 13
   value engineering 224
Contracts
   A/E 60
   AIA 119
   addenda 96
   advantage of writing own 111
   advantages of two contracts..........55
   agency CM 34
   arbitration requirement 244
   bias....................................... 116-117
   CM at risk.....................................20
   CM contractor 13
   CM scope of work................... 54, 57
   complying with federal law 119
   complying with state law 118
   Consensus DOCS.........................119
   construction 52-54
   Construction Contract Writer............56
   construction management...........112
   cost plus 10 percent 219
   DBIA 119
   design termination clause..............61
   differing site conditions.........236-237
   download from Web....................112
   download from website.................20
   drafting.......................................112
   EJCDC..119
   essential elements 112-113
   Federal Acquisition Regulations...237
   grounds for termination 211
   illegal 118-119
   importance in court 24, 30-31

incorporation by reference 61
insurance requirements 127
intent of language 200
interpretation 199-200
language 200
laws regarding 118-119
lease-lease 88
legitimate delays 139-140
liquidated damages clause 144
mediation requirement 241
model 119-120
no-damage-for-delay provision ...143
no-lien contract 188
optional contract details 114-115
pre-construction 51, 64
record documents 168
terms that shift bias 117
three-day right of rescission 119
trade 112
value engineering protection 225
website 118
Correspondence file 163
Cost
brand substitutions 72
change orders 215
owner delay 142
status reports 159
Cost estimates
contractor 77-78
owner's budget 77
published 79
Cost plus contracts, cardinal
change doctrine 219
Court case, Thurber Lumber v.
Marcario 23-31, 34
Courts
deciding disputes 245
good records win disputes 164
Critical path method 84
CxA 145

D

Damage, contractor caused 186
DBIA model contracts 119
Debris removal, site 75
Default insurance 131
Defect bond 130
Defective work 147-148
grounds for nonpayment 185
Delays 84-85
act of God 139-140
avoiding claims 143-144
cost to owner 155
force majeure 139-140
owner caused 140-143
weather 220
Deliveries, inspecting 166
Design
A/E process 61-62
CM involvement 38-39
cost estimates 61
defects, correcting 205-206
drafting problems 73
phases 62-63, 67
reducing risk of claims 227
selecting A/E team 57-60

value engineering 226
weekly cost estimates 77
Design-Build Institute of America
(DBIA) 119
Details, plan 73
Differing site conditions 236-237
forces of nature 238
man-made 238
Dimensions, design issues 73
Directory, project 167
Disclosures, site 235
Discretionary suspension 209-210
Dispute resolution
bonds 130
contract details 115
Disputes
arbitration 244
CM/owner, avoiding 154-155
contractor/owner, avoiding 151
litigation 245
mediating 241
resolving 239-240
retain good records 163-164
Documents, contract 96
Drafting programs, dimension
issues 73
Drainage, site 76

E

Early contractor involvement
(ECI) 37-38
Early occupancy
acceptance of work 248
insurance liability 248
multiple warranty periods 249
ECI contractor 37-38
EJCDC model contracts 119
Electrical work, plan check 74
Emergency contact list 167
Engineering, selecting A/E
team 57-60
Engineers Joint Contract
Documents Committee (EJCDC) ...119
Equal materials 72
Errors and omissions
insurance, architects 232
plans 68, 70-77
Estimates
contractor cost estimate 77-78
ROM 221
Evaluating payment requests 171
Excavation, pre-bid conference 100
Exclusions, warranty 261
Explosion insurance 128
Explosives on-site 150
Express warranties 259
Extra work
claims 234
contract details 114
differing site conditions 237

F

Feasibility study, cost estimate 78
Federal Acquisition Regulations 237
Federal contract law 119

Fee
building permit 81
CM contractor 14, 17-19
plan check 81
Final completion, sign off
punch list 256
Final payment 173
lien waiver 188
release of claims 178-179
substantial completion 177-178
Fixed cost contracts, tracking
payments 191
Flame-spread material class 199
Flammable liquids 150
Float time 84-85
in schedule 135
Force account 215
Force majeure delays 139-140
Forms
Bid Form 107-108
Change Order Agreement 222
Invitation to Bid 103-110
Pre-Qualification Statement 91-94
Purchase Order Cover Sheet 122
Request for Quotation Cover
Sheet 123
Rules for Bidding 109-110

G

General contractor
bids 87
bonding requirements 127
change order claims 224
claims 232-233
claims for delay 140-143
claims for extra work 237-238
contract terms that shift bias 117
compensation for early
occupancy 248
cost estimate 77-78
defective work 148
ECI 37-38
final completion invoice 256
hazardous materials 149
insurance requirements 127
jobsite safety 146
lien rights 187
preliminary punch list 250
pricing change orders 215-216
project development 38-39
qualifying for bid 89-90
reporting plan discrepancies 236
role in construction 125-127
statutory employer 27
termination for cause 211
traditional role 7, 13
value engineering 224
Green, building 144-145
reducing labor and materials 228
Guaranteed maximum price 20

H

Hazardous materials 149-150
contract details 114
Hourly rate, CM 17

I

Illegal contracts 118-119
Implied warranty 15, 258-259
Incorporation by reference 61
Independent commissioning
    authority (CxA) 145
Information
    channeling 154, 156-158
    CM 157-158
    contact list 167
    correspondence file.......................163
    job log 165-166
    legal authority for decisions........155
    money source for project 155
    owner-supplied materials 156
    submittal log............................160-161
Initial payments..................................173
Inspections
    contract details 115
    schedules................................ 134-135
    special 76
Inspector, auditing payments..........190
Installation compliance 199
Insurance
    contract details 113
    contract requirements 127
    types of coverage 128-129
Interpretation
    code...204
    contract 199-200
    plans and specs 201-202
Invitation to Bid
    addenda 96
    forms.. 103-110
    plan approval 97
    preparing documents................ 95-96
Invoice payments
    evaluating payment requests......172
    grounds for rejection 184-187
    recommend payment 182

J

Jack the pool builder..................... 47-54
Job conference
    chain of authority 207
    discussion topics...........................208
Job description, CM contractor 10
Job log 165-166
Job problems, status reports 159
Job sequences, scheduling 84
Job suspension...................................209
    contract details 115
Job walk
    avoid claims......................... 233-234
    pre-bid conference 97
Jobsite
    cleanup 149
    differing conditions 236
    safety 145-146

L

Labor and materials
    conserving......................................228
    estimating 79
Language, contract............................200
Law, contract................................118-119
Lead-based paint................................150
Leadership in Energy &
    Environmental Design (LEED)......144
Lease-lease contracts 88
LEED certification 144-145
    recycling construction debris......149
Legal force, building process...........204
Lender, project inception estimate....78
Letters of transmittal 160
Liability insurance 128
License, CM contractor................. 14-15
Liens
    contract details 113
    release at final completion...........179
    waivers 187
Limitations, site 76
Liquidated damages 144
    exceed payment due.....................185
    late completion..............................239
Litigation, claims and disputes........245
Little Miller Act 187
Loss, mitigating.......................... 230-231
Low bid...................................... 100-101
Lump sum, CM contractor fee 18

M

Maintenance bond 130
Manufacturer, installation
    compliance 199
Markup
    change orders 219
    CM contract 16
Material safety data sheets
    (MSDS)..149
Materials
    change order for 218
    flame-spread class.......................199
    ordered by owner 156
    plan review 72-73
    published standards............. 198-199
    stored off-site 181
Means and methods of
    construction..................................129
Mechanic's lien 56
    contract details 113
    release at final completion...........179
    waivers 187
Mechanical work, plan check............75
Mediation
    finding common ground.............243
    informal 151
    locating a mediator......................242
    resolving claims 241
Miller Act 130
    payment bond 187
Model contracts 119-120
Money, know source 155
Monthly charge, CM...........................18
MSDS 149

N

Narrow interpretation, plans 202
National Fire Protection
    Association (NFPA).......................124
National standards testing, ASTM...199
Negotiated change orders................219
Negotiated contract awards 88-89
NFPA *National Electrical Code* 124
No-lien contract..................................188
Notice of claim...................................229

O

Occupancy
    beneficial 247
    requirements.................................248
Off-site storage, materials................181
Omissions, errors and
    insurance, architects 232
    plans 68, 70-77
Open bidding............................... 88-89
    construction publication...............95
Overhead and profit, CM contract ...16
Owner
    acknowledge substantial
        completion 251
    beneficial occupancy 247
    contract terms that shift bias....... 117
    contract with CM 24-25, 30-31
    delays, contractor claims 140-143
    design termination 61
    explain rejected invoices to 184
    final completion payment ...178, 256
    hazardous materials on-site 149
    insurance requirements 127
    jobsite safety 146
    keeping informed 153
    legal authority for decisions........155
    lien waivers...................................187
    need for CM contractor...................8
    pay on CM recommendation 172
    project development................ 38-39
    provide with status reports 159
    record documents 168
    responsibilities 197
    site cleanup 150
    terminate job for cause..........211-212
Owner-controlled wrap-up
    insurance 129

P

Paper Contracting website 20, 112
Paper contractor
    defined... 5-6
    responsibilities 5-7
Paper trail, payment process 172
Partial completion..................... 253-254
Pay, CM contractor...................... 17-19
Pay-when-paid clause 188-189
Payment bond....................................130
Payment process
    evaluating requests......................171
    final completion payment 256
    initial payment 173

interest on past due balance 181
materials stored off-site 181-182
progress payments 173
schedule of values 175
T & M contracts 190
tracking payments 189
Payments, contract details 113
Percent of job cost, CM
contractor fee 18-19
Performance bond 130
Performance specs 73, 205
Permit bond 130
Permits
applying for 79-80
contract details 114
fees 81
posting 82
PERT scheduling 84
Plan check fees 81
Plan checking
anticipate problems 68
guidelines 69
Plan review
building department checklist 81
check for inconsistencies 74
check plan notes 74
cost issues 72
drafting problems 73
during design phase 67-68
electrical work 74
errors and omissions 77
mechanical work 75
permit issuing process 80-82
plan details 71
resolving issues 77
safety requirements 76
schedules 74
site conditions 75
six C's to check for 70-72
special inspections 76
testing requirements 76
warranty protection 77
Plan rooms 90
Planning department, permit
application 80-82
Plans
addenda 96
architect's intent 204
digital 90
errors in 235
incomplete 206-207
paper 95
reporting discrepancies 236
rules of interpretation 201-202
Plans examiner 81-82
Prayer meeting 41-42
Pre-bid conference 207-208
address scheduling issues 100
address site issues 98-100
job walk 97
Pre-bid site walk, avoiding
claims 233-234

Pre-construction
CM contractor services 37
CM contract 51
contracts, CCW 64
sequence 40
Pre-Qualification Statement 91-94
Pre-qualifying, bidders 89-90
Preliminary design 63
estimate 78
Pricing change orders 215-216
time and material cost 217
Problems, job, status reports 159
Product liability 128
Product testing, SDOs 198
Program manager 7
Programming, architectural 62
Progress payments 175
by job phase 174
final 256
initial 173
schedule of values 175
Project
awarding contract 100-101
bids 95
closeout 247
CM contractor role 10, 38-39
conserve labor and materials 228
construction surprises 234
contact directory 167
contract details 112-113
delays in schedule 84-85
differing site conditions 236
early occupancy 248
estimate 78-79
final project report 257
LEED 144-145
legal force behind 204
legitimate delays 139-140
oversight by CM 126
partial completion 253-254
payments 173-175, 256
plans vs. specs 202-203
record documents 168
retaining records 163
schedule 82
scheduling 132, 134-135
status reports 159
substantial completion 249
suspension 209
termination for cause 211
walk-through inspection 250-252
Project closeout, record
documents 168
Public works projects
bid competition 87, 89
bid regulations 95
lease-lease contracts 88
Miller Act 187
Punch list
before occupancy 249
completing 255
preliminary 250
sign off at final completion 256
substantial completion 254-255
Purchase Order Cover Sheet 122
Purchase orders 121
inspection and testing
compliance 124

Q

Quality assurance program 148
Quote, supplier 121-124

R

Record documents 168
Record retention 163
Records, maintaining 64
Recycling construction debris 145
jobsite cleanup 149
Reg Z notice 56
Rejecting invoice payments 183
explain to owner 184
grounds for rejection 184-187
Release of claims 178-179
Request for Quotation Cover
Sheet 123
Residential project, CM
contractor role 10-11
Resources, using efficiently 144-145
Retainage
release of 174, 179-180
subcontractor 181
Retainer 173
Risk
general contracting 8-9
less for CM contractor 17
paper contracting 6
Risk allocation 228
Rough order of magnitude
(ROM) estimate 221
Rules for Bidding 109-110
Rules of interpretation, plans ... 201-202

S

Safety
contract details 114
emergency contact list 167
equipment required 146
jobsite 145
owner's responsibilities 146
pre-bid conference 100
program 146
site conditions 76
Sales, successful approach 50-51
Samples
detail on job schedule 136
tracking submittals 161
Schedule
approving 136-137
components 84
details 134-135
float time 84-85
importance of 83
job sequences 84
keep job moving 137-139
plan check 74
project 82
software 83
status reports 159
track dates/deadlines 136

Schedule of values
    approving 176-177
    payment schedule 175
    tracking payments 191
Schematic design 63
Scope of work
    contract details 113
    modify with change order 213
Security, pre-bid conference 100
Shop drawings
    detail on job schedule 136
    tracking submittals 161
Site conditions
    differing 236
    limitations 76
    plan check 75
    soils 99
Site issues, pre-bid conference ... 98-100
Site planning, contract details 114
Site walk, avoiding claims 233-234
Small claims court 245
Software, scheduling 83
Soils, pre-bid conference 99
Specs
    performance vs. brand name 205
    rules of interpretation 201-202
Spreadsheet, payment
    tracking 190-192
Square footage estimate 78
Standard contracts
    A/E 60
    bias 116
Standards development
    organization (SDO) 198
Standards, testing organizations 124
State contract law 118
State law
    liens 187
    payment process 171
    rejecting invoice payments 184
Status reports 159
Statutory employer 27
Subcontractors
    contract details 114
    contract with owner 13
    lien rights 187
    payments 188-189
    retainage 181
    scheduling 84, 134-135
Submittals
    architects review 162
    detail on job schedule 136
    keep log 160
    samples 161
    shop drawings 161
    substitutions 161
Substantial completion 249
    acknowledgment memo 255
    contract price due 251
    punch list 254-255
    walk through 177-178
Substitution submittal 161
Suppliers
    dealing with CM 26
    invoice owners 31-32
    lien rights 187
    payment 188
    quotes 121-124
Surety bond 130
Suspension for cause 209

T

T & M contracts, payment
    tracking 190
Takeoff, labor and materials 79
Tasks, scheduling 84, 134-135
Tenant Improvement project,
    CM contractor role 10, 12
Termination for cause 211
Testing
    materials 198-199
    requirements 76
    standards organizations 124
Three-day notice of right to cancel ... 56
Three-day right of rescission 119
Thurber Lumber v. Marcario .. 23-31, 34
Tracking payments 190
Trade, schedules 134-135
Trade contractors
    contract with owner 13
    insurance requirements 127
    lien rights 187
    negotiating contracts 87
    safety equipment 146
    scheduling 84
    working for owner 31-32
Trade contracts 112
    complying with state law 119
Traffic plan, jobsite 76
Truth in Lending Act 119

U

U.S. Green Building
    Certification Institute 144-145
Underground hazards liability
    insurance 128
Unforeseeable conditions 238

V

Value engineering
    contractor proposal 224-225
    during design phase 226
    protection of idea 225
Vehicle insurance 129

W

Waivers
    contract details 113
    lien 187
Walk-through inspection 178
    serious defects 252
    substantial completion 249-250
Warranties
    begin at occupancy 249
    callback period 255
    claims 262-263
    construction defects 258
    contract details 115
    exclusions 261
    express 259-260
    implied 258-259
    record documents 168
Warranty law, construction 15-16
Warranty protection 77
Websites
    construction contract 118
    contract downloads 112
    Paper Contracting 20, 112
Weekly charge, CM 18
Weekly estimates, design phase 78
Work, defective 147-148
Worker's compensation
    insurance 128
    required 27, 33
Working drawings 63
Workmanlike construction 16
Wrap-up insurance policy 129
Written agreements,
    importance of 24-25, 30-31

X

X C U liability coverage 128

Practical References for Builders

National Construction Estimator

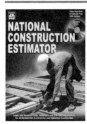

Current building costs for residential, commercial, and industrial construction. Estimated prices for every common building material. Provides manhours, recommended crew, and gives the labor cost for installation. Includes a CD-ROM with an electronic version of the book with *National Estimator*, a stand-alone *Windows*™ estimating program, plus an interactive multimedia video that shows how to use the disk to compile construction cost estimates.
672 pages, 8½ x 11, $72.50. Revised annually

2009 *International Residential Code*

Replacing the *CABO One- and Two-Family Dwelling Code*, this book has the latest technological advances in building design and construction. Among the changes are provisions for steel framing and energy savings. Also contains mechanical, fuel gas and plumbing provisions that coordinate with the *International Mechanical Code* and *International Plumbing Code*. **868 pages, 8½ x 11, $88.00**

Also available:
2006 *International Residential Code* $81.50
2003 *International Residential Code*, $72.50
2000 *International Residential Code*, $59.00
2000 *International Residential Code* on CD-ROM, $48.00

Construction Forms for Contractors

This guide contains 78 practical forms, letters and checklists, guaranteed to help you streamline your office, organize your jobsites, gather and organize records and documents, keep a handle on your subs, reduce estimating errors, administer change orders and lien issues, monitor crew productivity, track your equipment use, and more. Includes accounting forms, change order forms, forms for customers, estimating forms, field work forms, HR forms, lien forms, office forms, bids and proposals, subcontracts, and more. All are also on the CD-ROM included, in Excel spreadsheets, as formatted Rich Text that you can fill out on your computer, and as PDFs.
360 pages, 8½ x 11, $48.50

Insurance Restoration Contracting: Startup to Success

Insurance restoration — the repair of buildings damaged by water, fire, smoke, storms, vandalism and other disasters — is an exciting field of construction that provides lucrative work that's immune to economic downturns. And, with insurance companies funding the repairs, your payment is virtually guaranteed. But this type of work requires special knowledge and equipment, and that's what you'll learn about in this book. It covers fire repairs and smoke damage, water losses and specialized drying methods, mold remediation, content restoration, even damage to mobile and manufactured homes. You'll also find information on equipment needs, training classes, estimating books and software, and how restoration leads to lucrative remodeling jobs. It covers all you need to know to start and succeed as the restoration contractor that both homeowners and insurance companies call on first for the best jobs.
640 pages, 8½ x 11, $69.00

Markup & Profit: A Contractor's Guide, Revisited

In order to succeed in a construction business, you have to be able to price your jobs to cover all labor, material and overhead expenses, and make a decent profit. But calculating markup is only part of the picture. If you're going to beat the odds and stay in business — profitably, you also need to know how to write good contracts, manage your crews, work with subcontractors and collect on your work. This book covers the business basics of running a construction company, whether you're a general or specialty contractor working in remodeling, new construction or commercial work. The principles outlined here apply to all construction-related businesses. You'll find tried and tested formulas to guarantee profits, with step-by-step instructions and easy-to-follow examples to help you learn how to operate your business successfully. Includes a link to free downloads of blank forms and checklists used in this book.
312 pages, 8½ x 11, $47.50

CD Estimator

If your computer has *Windows*™ and a CD-ROM drive, CD Estimator puts at your fingertips over 150,000 construction costs for new construction, remodeling, renovation & insurance repair, home improvement, framing & finish carpentry, electrical, concrete & masonry, painting, earthwork & heavy equipment and plumbing & HVAC. Quarterly cost updates are available at no charge on the Internet. You'll also have the *National Estimator* program — a stand-alone estimating program for *Windows*™ that *Remodeling* magazine called a "computer wiz," and *Job Cost Wizard*, a program that lets you export your estimates to *QuickBooks Pro* for actual job costing. A 60-minute interactive video teaches you how to use this CD-ROM to estimate construction costs. And to top it off, to help you create professional-looking estimates, the disk includes over 40 construction estimating and bidding forms in a format that's perfect for nearly any *Windows*™ word processing or spreadsheet program.
CD Estimator is $108.50

Construction Contract Writer

Relying on a "one-size-fits-all" boilerplate construction contract to fit your jobs can be dangerous — almost as dangerous as a handshake agreement. Construction Contract Writer lets you draft a contract in minutes that precisely fits your needs and the particular job, and meets both state and federal requirements. You just answer a series of questions — like an interview — to construct a legal contract for each project you take on. Anticipate where disputes could arise and settle them in the contract before they happen. Include the warranty protection you intend, the payment schedule, and create subcontracts from the prime contract by just clicking a box. Includes a feedback button to an attorney on the Craftsman staff to help should you get stumped — No extra charge. **$99.95**. Download the Construction Contract Writer at:
http://www.constructioncontractwriter.com

Contractor's Guide to *QuickBooks Pro* 2010

This user-friendly manual walks you through *QuickBooks Pro*'s detailed setup procedure and explains step-by-step how to create a first-rate accounting system. You'll learn in days, rather than weeks, how to use *QuickBooks Pro* to get your contracting business organized, with simple, fast accounting procedures. On the CD included with the book you'll find a *QuickBooks Pro* file for a construction company. Open it, enter your own company's data, and add info on your suppliers and subs. You also get a complete estimating program, including a database, and a job costing program that lets you export your estimates to *QuickBooks Pro*. It even includes many useful construction forms to use in your business.
344 pages, 8½ x 11, $57.00
See checklist for other available editions.

Building Code Compliance for Contractors & Inspectors

Have you ever failed a construction inspection? Have you ever dealt with an inspector who has his own interpretation of the Code and forces you to comply with it? This new book explains what it takes to pass inspections under the 2009 *International Residential Code*. It includes a Code checklist — with explanations and the Code section number — for every trade, covering some of the most common reasons why inspectors reject residential work. The author uses his 30 years' experience as a building code official to provide you with little-known information on what code officials look for during inspections. **232 pages, 8½ x 11, $32.50**

Builder's Guide to Accounting Revised

Step-by-step, easy-to-follow guidelines for setting up and maintaining records for your building business. This practical guide to all accounting methods shows how to meet state and federal accounting requirements, explains the new depreciation rules, and describes how the Tax Reform Act can affect the way you keep records. Full of charts, diagrams, simple directions and examples to help you keep track of where your money is going. Recommended reading for many state contractor's exams. Each chapter ends with a set of test questions, and a CD-ROM included FREE has all the questions in interactive self-test software. Use the Study Mode to make studying for the exam much easier, and Exam Mode to practice your skills. **360 pages, 8½ x 11, $35.50**

How to Succeed With Your Own Construction Business

Everything you need to start your own construction business: setting up the paperwork, finding the jobs, advertising, using contracts, dealing with lenders, estimating, scheduling, finding and keeping good employees, keeping the books, and coping with success. If you're considering starting your own construction business, all the knowledge, tips, and blank forms you need are here. **336 pages, 8½ x 11, $28.50**

Contractor's Plain-English Legal Guide

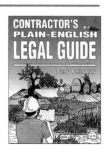

For today's contractors, legal problems are like snakes in the swamp — you might not see them, but you know they're there. This book tells you where the snakes are hiding and directs you to the safe path. With the directions in this easy-to-read handbook you're less likely to need a $200-an-hour lawyer. Includes simple directions for starting your business, writing contracts that cover just about any eventuality, collecting what's owed you, filing liens, protecting yourself from unethical subcontractors, and more. For about the price of 15 minutes in a lawyer's office, you'll have a guide that will make many of those visits unnecessary. Includes a CD-ROM with blank copies of all the forms and contracts in the book. **272 pages, 8½ x 11, $49.50**

Estimating Home Building Costs Revised

Accurate estimates are the foundation of a successful construction business. Leave an item out of your original estimate and it can take the profit out of your entire job. This practical guide to estimating home construction costs has been updated with Excel estimating forms and worksheets with active cells that ensure accurate and complete estimates for your residential projects. Load the enclosed CD-ROM into your computer and create your own estimate as you follow along with the step-by-step techniques in this book. Clear, simple instructions show how to estimate labor and material costs for each stage of construction, from site clearing to figuring your markup and profit. Every chapter includes a sample cost estimate worksheet that lists all the materials to be estimated. Even shows how to figure your markup and profit to arrive at a price. **336 pages, 8½ x 11, $38.00**

 Craftsman Book Company
6058 Corte del Cedro
P.O. Box 6500
Carlsbad, CA 92018

☎ **24 hour order line**
1-800-829-8123
Fax (760) 438-0398

In A Hurry?

We accept phone orders charged to your

○ Visa, ○ MasterCard, ○ Discover or ○ American Express

Card#_____

Exp. date_____ Initials_____

Tax Deductible: Treasury regulations make these references tax deductible when used in your work. Save the canceled check or charge card statement as your receipt.

Name_____

e-mail address (for order tracking and special offers)

Company_____

Address_____

City/State/Zip   ○ This is a residence

Total enclosed _____ (In California add 7.25% tax)
We pay shipping when your check covers your order in full.

Order online www.craftsman-book.com
Free on the Internet! Download any of Craftsman's estimating databases for a 30-day free trial!
www.craftsman-book.com/downloads

Download all of Craftsman's most popular costbooks for one low price with the Craftsman Site License: www.craftsmansitelicense.com

10-Day Money Back Guarantee	Prices subject to change without notice
○ 35.50 Builder's Guide to Accounting Revised	○ 38.00 Estimating Home Building Costs Revised
○ 32.50 Building Code Compliance for Contractors & Inspectors	○ 28.50 How to Succeed With Your Own Construction Business
○ 108.50 CD Estimator	○ 69.00 Insurance Restoration Contracting: Startup to Success
○ 48.50 Construction Forms for Contractors	○ 88.00 2009 *International Residential Code*
○ 57.00 Contractor's Guide to *QuickBooks Pro* 2010	○ 81.50 2006 *International Residential Code*
○ 56.50 Contractor's Guide to *QuickBooks Pro* 2009	○ 72.50 2003 *International Residential Code*
○ 54.75 Contractor's Guide to *QuickBooks Pro* 2008	○ 59.00 2000 *International Residential Code*
○ 53.00 Contractor's Guide to *QuickBooks Pro* 2007	○ 48.00 2000 *International Residential Code* on CD-ROM
○ 49.75 Contractor's Guide to *QuickBooks Pro* 2005	○ 47.50 Markup & Profit: A Contractor's Guide, Revisited
○ 48.50 Contractor's Guide to *QuickBooks Pro* 2004	○ 72.50 National Construction Estimator with FREE *National Estimator* on a CD-ROM
○ 47.75 Contractor's Guide to *QuickBooks Pro* 2003	
○ 45.25 Contractor's Guide to *QuickBooks Pro* 2001	○ 55.50 Paper Contracting
○ 49.50 Contractor's Plain English Legal Guide	○ FREE Full Color Catalog

Download free construction contracts legal for your state: www.construction-contract.net

Craftsman Book Company
6058 Corte del Cedro
P.O. Box 6500
Carlsbad, CA 92018

☎ 24 hour order line
1-800-829-8123
Fax (760) 438-0398

In A Hurry?
We accept phone orders charged to your
❏ Visa, ❏ MasterCard, ❏ Discover or ❏ American Express

Name _____

e-mail address (for order tracking and special offers) _____

Company _____

Address _____

City/State/Zip _____ This is a residence

Total enclosed _____ (In California add 7.25% tax)
We pay shipping when your check covers your order in full.

Card # _____

Exp. date _____ Initials _____

Tax Deductible: Treasury regulations make these references tax deductible when used in your work. Save the canceled check or charge card statement as your receipt.

Order Online www.craftsman-book.com
Free on the Internet! Download any of Craftsman's estimating costbooks for a 30-day free trial! www.costbook.com

10-Day Money Back Guarantee / Prices Subject to Change Without Notice

- 35.50 Builder's Guide to Accounting Revised
- 32.50 Building Code Compliance for Contractors & Inspectors
- 108.50 CD Estimator
- 48.50 Construction Forms for Contractors
- 57.00 Contractor's Guide to QuickBooks Pro 2010
- 56.50 Contractor's Guide to QuickBooks Pro 2009
- 54.75 Contractor's Guide to QuickBooks Pro 2008
- 53.00 Contractor's Guide to QuickBooks Pro 2007
- 49.75 Contractor's Guide to QuickBooks Pro 2005
- 48.50 Contractor's Guide to QuickBooks Pro 2004
- 47.75 Contractor's Guide to QuickBooks Pro 2003
- 45.25 Contractor's Guide to QuickBooks Pro 2001
- 49.50 Contractor's Plain English Legal Guide
- 38.00 Estimating Home Building Costs Revised
- 28.50 How to Succeed With Your Own Construction Business
- 69.00 Insurance Restoration Contracting: Startup to Success
- 88.00 2009 *International Residential Code*
- 81.50 2006 *International Residential Code*
- 72.50 2003 *International Residential Code*
- 59.00 2000 *International Residential Code*
- 48.00 2000 *International Residential Code* on CD-ROM
- 47.50 Markup & Profit: A Contractor's Guide, Revisited
- 72.50 National Construction Estimator with FREE *National Estimator* on a CD-ROM
- 55.50 Paper Contracting
- FREE Full Color Catalog

Download all of Craftsman's most popular costbooks for one low price with the Craftsman Site License: www.craftsmansitelicense.com

Download a FREE trial Construction Contract Writer

Writing contracts that comply with the law in your state isn't easy. A contract that doesn't comply could leave you with no way to collect. You need contracts that:

❏ Fit your jobs exactly.
❏ Contain your state's required statutes and attachments.
❏ Will stand up in court should a disagreement occur.

No pre-made contract fits all jobs and state contract requirements. You'll be unprotected should a disagreement occur and you end up in court. In some states you can even lose your license.

Now there's a better way. Write perfect contracts in minutes by answering simple interview questions:

❏ No legal background needed.
❏ All state-required statutes, attachments and warranties included.
❏ You create contracts as detailed or as simple as the job and your state requires.

If you get stuck, click the *"Get Help From an Attorney"* button. You'll have an answer in 24 hours. No charge. No limit.

Take control of the contract-drafting process and you protect yourself, as well as control the bottom line.

The trial download is free. The full download is **$99.95**. Updates for your state are free for one year.

Get Construction Contract Writer for your state at: www.craftsman-book.com

Mail This Card Today
for a Free Full-Color Catalog

Over 100 books, annual cost guides and estimating software packages at your fingertips, with information that can save you time and money. Here you'll find information on carpentry, contracting, estimating, remodeling, electrical work and plumbing.

All items come with an unconditional 10-day money-back guarantee. If they don't save you money, mail them back for a full refund.

Name _____

e-mail address (for special offers) _____

Company _____

Address _____

City/State/Zip _____

Craftsman Book Company / 6058 Corte del Cedro / P.O. Box 6500 / Carlsbad, CA 92018

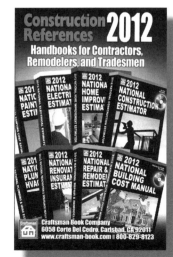

BUSINESS REPLY MAIL

FIRST CLASS MAIL PERMIT NO. 271   CARLSBAD, CA

POSTAGE WILL BE PAID BY ADDRESSEE

 Craftsman Book Company
6058 Corte del Cedro
P.O. Box 6500
Carlsbad, CA 92018-9974

NO POSTAGE
NECESSARY
IF MAILED
IN THE
UNITED STATES

BUSINESS REPLY MAIL

FIRST CLASS MAIL PERMIT NO. 271   CARLSBAD, CA

POSTAGE WILL BE PAID BY ADDRESSEE

 Craftsman Book Company
6058 Corte del Cedro
P.O. Box 6500
Carlsbad, CA 92018-9974

NO POSTAGE
NECESSARY
IF MAILED
IN THE
UNITED STATES

BUSINESS REPLY MAIL

FIRST CLASS MAIL PERMIT NO. 271   CARLSBAD, CA

POSTAGE WILL BE PAID BY ADDRESSEE

 Craftsman Book Company
6058 Corte del Cedro
P.O. Box 6500
Carlsbad, CA 92018-9974

NO POSTAGE
NECESSARY
IF MAILED
IN THE
UNITED STATES